ADVANCE PRAISE FOR

THE MYTHS of SAFE PESTICIDES

"André Leu breaks the chain of corporate misinformation on pesticides and unmasks the failure of chemical warfare against pests. As Leu points out, there are ecologically, economically, and socially viable alternatives to chemical agriculture."

Miguel A. Altieri, PhD, professor of agroecology, University of California–Berkeley

"A must-read! All that you didn't know about pesticides will knock your socks off. But don't worry, Leu's powerful, fast-moving account of our pesticide-planet will charge you up, not wipe you out. Never again feel tongue-tied when explaining why eating organic is more than worth it. Leu's pulled it all together for us, brilliantly. Fascinating reading, solidly backed by science. Plus, what Leu knows firsthand, and proves here with evidence, is that we can feed ourselves well—*all* of us—without pesticides' horrific harms."

Frances Moore Lappé, author, Diet for a Small Planet, *cofounder, Small Planet Institute*

"Piercing, commonsensical details that dissolve defective, manipulative science designed solely by the poisoners. Enjoy this empowerment and help stop criminal profits. If you do not know, you will not protect yourself or those you should."

Michael Potter, president, 44 years at Eden Foods

"André Leu does a phenomenal job sharing over forty years of organic farming experience, and the issues surrounding pesticides. This book provides evidence from hundreds of published scientific studies that will likely cause anyone currently using pesticides to reconsider their approach and seriously question how they can be called safe. Highly recommended."

Dr. Joseph Mercola, founder, Mercola.com

"As a staunch opponent of pesticides, I thought I knew the arguments. In this wonderful gift to the integrity food movement, André Leu gives even the faithful a quotable, understandable, captivating text to affirm and strengthen our arguments."

Joel Salatin, founder, Polyface Farms, author, featured in The Omnivore's Dilemma *and* Food, Inc.

"If you already understand your role in the ecosystem by carefully choosing your food, André Leu, in an engaging style, will deepen your passion. If you are new to looking at science that debunks the claims of agrochemical safety and judicious regulatory oversight, this book will have the hair standing up on the back of your neck."

Mark A. Kastel, codirector, senior farm policy analyst, The Cornucopia Institute

"Read this book if you need to remember or learn why the transition to organic practices on a massive scale should be among our highest priorities. *The Myths of Safe Pesticides* lays the groundwork for arguing the urgent need to embrace organic management practices that eliminate reliance on pesticides and the resulting severe or uncertain adverse impacts on health and the environment."

Jay Feldman, executive director, Beyond Pesticides

"Critics of organics claim there is no research, data, or evidence that toxic chemical landscaping and farming products present problems, or that organic systems work better. André Leu brilliantly details the well-documented evidence to the contrary in clear language that everyone can understand—even those people and institutions trying to protect the status quo."

Howard Garrett, author, The Organic Manual

"André Leu's prescient exposé on the myth of pesticide safety hits the ball out of the park, corroborating what scientists like Rachel Carson and Theo Colborn have been forecasting for decades. The safe-when-used-as-directed mantra can no longer be justifiably embraced by the academic, scientific, or regulatory communities. Bravo, Mr. Leu, for laying it out in spades."

Jerry Brunetti, author, The Farm as Ecosystem

THE

MYTHS

of Safe Pesticides

THE
MYTHS
of **Safe Pesticides**

by André Leu

Acres U.S.A.
Austin, Texas

The Myths of Safe Pesticides

Acres U.S.A.
P.O. Box 301209
Austin, Texas 78703 U.S.A.
512-892-4400 • fax 512-892-4448
info@acresusa.com • *www.acresusa.com*

Printed in the United States of America

Front cover photography © boggy22/iStock/Thinkstock
Back cover photography © Federico Rostagno/iStock/Thinkstock
Roundup® and Roundup Ready® crops are registered trademarks of the Monsanto corporation.

Publisher's Cataloging-in-Publication

Leu, André, 1953–
    The myths of safe pesticides / André Leu. Austin, TX, ACRES U.S.A., 2014
        xxiv, 142 pp., 23 cm.
    Includes bibliographical references, index, tables, and illustrations
    ISBN 978-1-60173-084-8 (print edition)
    ISBN 978-1-60173-085-5 (ebook edition, epub)
    ISBN 978-1-60173-086-2 (ebook edition, mobi)
    1. Pesticides—Health aspects. 2. Pesticide residues in food. 3. Spraying and dusting residues in agriculture. 4. Pesticides—Environmental aspects.
I. Leu, André, 1953– II. Title.

TD887.P45L48   2014   632.95042

To my wife, Julia, for supporting me for
almost forty years in all my various endeavours,
through tough times and good times.
I could not have achieved them without you.

To my sons, Asha and Nick. The twenty
years we had on our family farm together when
you were growing up were some of the most
wonderful in my life. I wish you the same joy
with your families.

All my love,
André

ACKNOWLEDGMENTS

My thanks to Dr. Nancy Swanson for all the graphs, the Rodale Institute and FiBL for the use of their photos, Dr. Elizabeth Guillete and *Environmental Health Perspectives* for the use of the Yaqui children's drawings, and Dr. Sue Edwards for the data on Tigray, Ethiopia. I particularly want to acknowledge my editor, Amanda Irle, for the work she has done to turn the basic text into a book that reads well, and Fred Walters, my publisher, for all his support.

CONTENTS

ABOUT THE AUTHOR

André Leu is the president of the International Federation of Organic Agricultural Movements (IFOAM), the world umbrella body for the organic sector. IFOAM has about 800 member organizations in 120 countries. Its goal is the worldwide adoption of ecologically, socially, and economically sound systems based on the principles of organic agriculture.

Kerry Larsen, Newsport

André has over forty years of experience in all areas of organic agriculture, including growing, pest control, weed management, marketing, post-harvest, transport, grower organizations, developing new crops, and education in Australia and in many other countries. He has an extensive knowledge of farming and environmental systems across Asia, Europe, the Americas, and Africa from forty years of visiting and working in these countries.

He has written and been published extensively in magazines, newspapers, journals, websites, and other media on many areas of organic agriculture, including climate change, the environment, and the health benefits of organic agronomy.

André and his wife own an agroecological organic tropical fruit orchard in Daintree, Queensland, Australia, that supplies quality-controlled fruit to a range of markets from local to international.

FOREWORD

The *Myths of Safe Pesticides* by André Leu takes us through nearly a century of pesticide use, from their introduction as poisons to kill humans in concentration camps and wars to their more recent function of killing pests in industrial chemical agriculture. It is no wonder that chemicals rooted in a militarized mind-set continue to cause harm today.

Rachel Carson, in chapter 3, "Elixirs of Death," of her book *Silent Spring*, indicates how the creation of pesticides was literally a by-product of war. "In the course of developing agents of chemical warfare, some of the chemicals created in the lab were found to be lethal to insects . . . some of them became deadly nerve gases. Others, of closely allied structure, became insecticides."

The Bhopal gas tragedy of 1984 is a stark reminder that pesticides kill. A gas leak from an Indian pesticide plant owned by Union Carbide, which is now owned by Dow, killed three thousand people in one night, and some estimates since count thirty thousand casualties. Pesticides in agriculture and food continue to kill farm workers, consumers, children, butterflies, bees, and others. The Bhopal tragedy is not over. Thousands continue to be harmed and maimed. And new Bhopals are being created in the country. In Kerala we have had the endosulfan tragedy. And Punjab has the infamous cancer train. Just in the last five years there have been 33,318 cancer deaths in Punjab. And the cancers are related to the heavy use of toxic chemicals in Green Revolution agriculture since 1965. These tragedies that have unfolded in Punjab and Bhopal are connected through the poisoned legacy of the Green Revolution, based on toxic chemicals—synthetic fertilizers and pesticides.

In our book, *Poisons in Our Food* (written by myself, Dr. Mira Shiva, and Dr. Vaibhav Singh and published in India by Natraj Publishers in 2012), we synthesize the research on the link between disease epidemics like cancer and the use of pesticides in agriculture in India. In *The Myths of Safe Pesticides*, André Leu wakes us to the global patterns of disease epidemics that are threatening human life and well-being with the combined threat from pesticides and pesticide-producing GMO plants.

As he reports, between 1980 and 2008 the number of people with type 2 diabetes has increased from 153 million to 347 million. The rate of autism in the United States was 1 in 88 children in 2012. By 2014 it had jumped to 1 in 68. Breast cancer, thyroid cancer, and urinary bladder cancer have increased exponentially in the United States with the increasing use of GMOs and glyphosate (Roundup).

Genetic engineering was offered as an alternative to chemical pesticides, but it is part of the same logic of the war against nature with poisons. Now the poison has been introduced as a toxin-producing gene into the plant, so the GMO becomes a pesticide-producing plant. Just as pesticides created pests instead of controlling them, GMOs as pesticide-producing plants increase pests and have created superpests instead of controlling them. New pests emerge, and old pests become resistant. The result is increased use of chemical pesticides.

Herbicide-resistant plants, such as Roundup Ready corn and soy, have led to an increased use of glyphosate, which kills all other plants including milkweed, the only type of plant that monarch butterflies use for laying their eggs. As Roundup Ready crops have increased to 90 percent of crops grown, milkweed has declined by 60 percent, and the number of monarch butterflies that migrate across the United States each year and overwinter in the forests of Mexico has dropped from 1 billion in 1997 to an all-time low of 33.5 million.

A recent study from Sri Lanka has shown that there is an epidemic of kidney failure related to the increasing use of glyphosate. Sri Lanka banned its use but reversed the ban after pressure from the agrochemical industry. Meanwhile, in the United States, Roundup is failing to control weeds and has led to the emergence of superweeds. GMOs are now being made resistant to an ingredient of Agent Orange.

Ecologically biodiverse systems do not just protect bees and pollinators that feed us. They control pests through pest-predator balance, supporting an abundance of natural enemies that prevent explosion of pest populations. Monocultures create a feast for pests, because there is no biodiversity to provide the ecological functions of pest control. However, in the industrial paradigm, pest control is an issue of war. As a pest-management textbook states, "The war against pests is a continuing one that man must fight to ensure his survival. Pests (in particular insects) are our major competitors on earth."

But the war against pests is neither necessary nor effective.

Pesticides create pests, they do not control them. Pests increase with the application of pesticides because beneficial species are killed, and pests become resistant. According to de Bach, "The philosophy of pest control by chemicals has been to achieve the highest kill possible, and percent mortality has been the main yardstick in the early screening of new chemicals in the lab. Such an objective, the highest kill possible, combined with ignorance of, or disregard for, nontarget insects and mites is guaranteed to be the quickest road to upset resurgences and the development of resistance to pesticides." Pests are controlled when there is ecological balance between diverse components of the farming system.

Biodiversity is our best friend in dealing with pest problems. First, pests do not emerge in agriculture systems based on diversity. Second, if there is a pest outbreak, biodiversity offers ecological alternatives, like botanical pest control agents like neem. The neem is a drought-resistant tree native to India that can be used as a natural alternative to synthetic pesticides. In 1984, at the time of the Bhopal disaster, I started a campaign, "No More Bhopals, Plant a Neem." Ten years later I found that the use of neem had been patented by the U.S. Department of Agriculture and W. R. Grace. With Magda Aelvoet from the Greens of European parliament and Linda Bullard, the former president of the International Federation of Organic Agriculture Movements, I filed a case challenging the biopiracy of neem. It took us eleven years, but we finally won, and the patent on using neem tree extracts as a fungicide was revoked.

A pest outbreak is a symptom of a system that is out of balance. What is needed is the reintroduction of balance through biodiversity. Instead, industrial agriculture deepens the imbalance by introducing more and more deadly poisons to kill pests. As Albert Howard observed, "The destruction of a pest is the evasion of, rather than the solution of, all agriculture problems."

As Rachel Carson concluded in *Silent Spring*:

> The "control over nature" is a phrase conceived in arrogance, born of the Neanderthal age of biology and philosophy, when it was supposed that nature exists for the convenience of man. The concepts and practices of applied entomology for the most part date from that Stone Age of science. It is our alarming misfortune that so primitive a science has armed itself with the most modern and terrible weapons, and that in turning them against the insects it has also turned against the earth.

A food and agriculture system based on biodiversity and free of chemicals and pesticides is the true answer to both pest control and to food and nutrition security. The real productivity of agriculture systems has not been measured because industrial agriculture focuses on commodity production, not food production. Monoculture commodity production requires intensive inputs of chemical fertilizers and pesticides; it requires intensive inputs of capital and fossil fuels. It uses ten kilocalories of energy as input to produce one kilocalorie of food as output. In terms of real productivity, it is a negative economy. But the illusion has been created that a food system that uses more inputs than it produces is efficient and productive, and we need it to expand.

More commodities mean less food. Only 10 percent of the corn and soy produced as a commodity is eaten by human beings. The rest goes to biofuel and animal feed. About 70 percent of the food eaten comes from small farms; only 30 percent comes from large industrial farms. While producing a smaller share of the global food basket, industrial chemical agriculture contributes to the larger share of the ecological problems of biodiversity erosion, water depletion and pollution, soil erosion and degradation, and climate change. We need to internalize these costs in the food system and not leave them as externalities to be born by society and other species.

We need to internalize health costs into the food and agriculture equation. Food is about nourishment and nutrition. Producing a nutritionally empty mass loaded with poisons is producing anti-food, not food.

If food is to provide nourishment instead of disease, we need a paradigm shift in agriculture. We need to be more specific when we refer to "intensification" of agriculture. The dominant system is intensive in terms of the use of pesticides, which have contributed to health hazards described in this book. It is intensive in monocultures, which are leading to the disappearance of biodiversity. With the destruction of the ecological functions of biodiversity in controlling pests and weeds, pesticide usage is leading to an increase in pest and weed problems, thus further increasing use of pesticides and poisons. It is also leading to an undermining of health and nutrition that can only come from ecologically sustainable biodiverse systems.

We need to move from chemical-intensive, fossil fuel–intensive, capital-intensive agriculture to biodiversity-intensive, ecologically intensive, knowledge-intensive agriculture based on the principles of agroecology and biodiversity. When measured in terms of biodiverse outputs and health and nutrition per acre, biodiverse ecological systems produce more food, and more nutrition, even as they control pests through ecological processes.

Not only does biodiversity provide ecological alternatives to pesticides, agroecological systems enhance farmers' incomes and thus contribute to reducing poverty.

Debt due to purchase of costly GMO seeds and pesticides has driven more than 284,000 farmers to suicide in India since 1995. No organic farmer using native seeds and cultivating biodiversity has committed suicide in India. All suicides are among farmers trapped in debt due to the costs of seeds and pesticides. Most suicides are concentrated in the cotton belt, which is now 95 percent Bt cotton. A pesticide-free, GMO-free agriculture can increase farmers' net income from two- to tenfold, which will translate into an agricultural system free of debt and suicide.

André Leu has presented us with the global synthesis of the scientific evidence of the harm to public health caused by pesticides and pesticide-producing GMOs. He has also provided us with the scientific evidence that pesticide-free alternatives are more productive.

Now is the time to make the transition and use our intelligence to farm without poisons, or in Rachel Carson's words, move beyond the "Neanderthal age of biology and philosophy," the age of pesticides and pesticide-producing plants.

Dr. Vandana Shiva
New Delhi, India, 2014

INTRODUCTION

A lifetime of farming with its joys, pleasures, and the immense satisfaction of literally harvesting the fruits of your labor also comes with the trials and tribulations of droughts, floods, pests, diseases, markets, and other uncertainties. Good farmers learn to closely observe, question, and study these challenges so that we can make better decisions on how to manage them. One of the reasons for this book is my observation of so much illness in our communities, especially cancers, behavioral disorders, and degenerative diseases. In my own case I would become ill every spraying season even though no sprays are used on my farm. The regulators and the extension officials would say that this was not related to all the pesticides used in agriculture because the science states that they are being used safely. Consequently, I decided to apply the lessons I'd learned as a farmer and, although I am not a scientist, I began to question this assertion by studying the published, peer-reviewed science around pesticides.

Conventional farming has grown dependent on using synthetic poisons as pesticides. These poisons are used in food production to eradicate pests, diseases, and weeds. The widespread use of these toxic substances is being justified by regulatory authorities as safe so long as they are used "correctly." Both industry and government are now promoting the concept of "good agricultural practice" to assure consumers that they do not need to have any concerns about the residues of numerous toxic pesticides in food. Some of these schemes are being certified as "safe food," "natural," "low pesticide," "environmental," etc. However, major international studies—such as the U.S. President's Cancer Panel 2010 report; the *International Assessment of Agricultural Knowledge, Science, and Technology for Development* report; *State of the Science of Endocrine Disrupting Chemicals 2012* by the World Health Organization and the United Nations Environment Programme; the United Nations Millennium Ecosystem Assessment Synthesis report; and many published studies by scientific researchers—have raised the issue of agricultural chemicals as significant contributors to negative global environmental change and adverse human health due to persistent and short-term environmental toxicants.[1]

> In 1999, Swiss research demonstrated that some of the rain falling on Europe contained such high levels of pesticides that it would be **ILLEGAL TO SUPPLY IT AS DRINKING WATER.**

The damage caused by agricultural chemicals in the environment and human health began to receive attention in the early 1960s when Rachel Carson wrote *Silent Spring*.[2] These chemicals were shown to persist and accumulate in the environment, causing mortality, birth defects, mutations, and diseases in humans and animals. The number and volume of chemicals used on our food and in the environment has increased exponentially since then.

In the 1990s the issue of chemicals disrupting the reproduction and hormone systems of all species, including humans, was brought to the public's attention by books like Theo Colborn, Dianne Dumanoski, and John Peterson Myers's *Our Stolen Future* and Deborah Cadbury's *The Feminization of Nature*. The peer-reviewed science summarized in these books showed that many chemicals, especially agricultural chemicals, were mimicking hormones such as estrogen, causing serious declines in fertility by reducing the quantity and quality of sperm production and damaging the genital urinary systems. They were major contributors to the dramatic rise in cancers of the sexual tissues—breast, uterine, ovarian, vaginal, testicular, and prostate cancers.[3]

The body of science showing that agricultural chemicals are responsible for declines in biodiversity along with environmental and health problems continues to grow. These toxic chemicals now pervade the whole planet, polluting our water, soil, air, and most significantly the tissues of many living organisms.[4] In 1999, Swiss research demonstrated that some of the rain falling on Europe contained such high levels of pesticides that it would be illegal to supply it as drinking water.[5] Rain over Europe was laced with atrazine, alochlor, 2,4-D, and other common agricultural chemicals sprayed onto crops. A 1999 study of rainfall in Greece found one or more pesticides in 90 percent of 205 samples taken.[6]

That inadequate pesticide regulation is resulting in major environmental and human health problems has been validated by hundreds of scientific studies. One of the most significant has been the 2010 report by the U.S. President's Cancer Panel (USPCP). This report was written by eminent scientists and medical specialists in this field, and it clearly states that environmental toxins, including chemicals used in farming, are the main causes of cancers. Published by the U.S. Department of Health and Human Services, the National Institutes of Health, and the National Cancer Institute, the report discusses many critical issues of chemical regulation.

> Nearly 1,400 pesticides have been registered (i.e., approved) by the Environmental Protection Agency (EPA) for agricultural and non-agricultural use. Exposure to these chemicals has been linked to brain/central nervous system (CNS),

breast, colon, lung, ovarian (female spouses), pancreatic, kidney, testicular, and stomach cancers, as well as Hodgkin and non-Hodgkin lymphoma, multiple myeloma, and soft tissue sarcoma. Pesticide-exposed farmers, pesticide applicators, crop duster pilots, and manufacturers also have been found to have elevated rates of prostate cancer, melanoma, other skin cancers, and cancer of the lip.

Approximately 40 chemicals classified by the International Agency for Research on Cancer (IARC) as known, probable, or possible human carcinogens, are used in EPA-registered pesticides now on the market.[7]

THE MYTHS

This book builds up a compelling body of evidence against pesticides based on hundreds of published scientific studies that seriously question the safety issues around the regulation of toxic chemicals. It proposes that much of the criteria used to underpin the current pesticide use in our food and environment are based on outdated methodologies rather than on the latest published science. Until the use of pesticides is regulated on the basis of current, published, peer-reviewed science, there is no sound scientific basis on which to base the belief that the residue levels in our food and environment are safe. Regulatory authorities are using data-free assumptions to perpetuate a series of mythologies to lull the public into a false sense of security about the safety of their levels of exposure to pesticides.

Given that there are thousands of chemical formulations used in the production of our food, this book would be too long if it went into detail for them all. Instead, I chose to highlight some of the most common agricultural chemicals as examples of the range of issues that surround the widespread use of these substances in our food supply and the environment. Many of the examples featured here are from the United States and Australia, as these are the countries that I know the best when it comes to pesticide use; however, the issues are similar in nearly every country.

The word "pesticide" is used in this book as a generic term for the numerous biocides that are used in agriculture, such as herbicides, fungicides, and insecticides. This book is focused primarily on the adverse health effects of pesticides on humans with many references to their adverse effects on other species as well. Data on the adverse effects of pesticides on the environment could fill another, even longer book as they are substantial and pervade every part of our planet.

NOTES

[1] Millennium Ecosystem Assessment Synthesis Report, United Nations Environment Programme, March 2005.

[2] Rachel Carson, *Silent Spring* (New York: Penguin Books, 1962).

[3] Theo Colborn, Dianne Dumanoski, and John Peterson Myers, *Our Stolen Future: Are We Threatening Our Fertility, Intelligence, and Survival? A Scientific Detective Story* (New York: Dutton, 1996); Deborah Cadbury, *The Feminization of Nature: Our Future at Risk* (Middlesex, England: Penguin Books, 1998).

[4] Kate Short, *Quick Poison, Slow Poison: Pesticide Risk in the Lucky Country* (St. Albans, NSW: K. Short, 1994); "U.S. President's Cancer Panel 2008–2009 Annual Report; Reducing Environmental Cancer Risk: What We Can Do Now," Suzanne H. Reuben for the President's Cancer Panel, U.S. Department Of Health And Human Services, National Institutes of Health, National Cancer Institute, April 2010; Åke Bergman et al., eds., *State of the Science of Endocrine Disrupting Chemicals 2012*, United Nations Environment Programme and the World Health Organization, 2013; Colborn, Dumanoski, and Myers, *Our Stolen Future*; Cadbury, *Feminization of Nature*.

[5] Fred Pearce and Debora Mackenzie, "It's Raining Pesticides," *New Scientist*, April 3, 1999, 23.

[6] Emmanouil Charizopoulos and Euphemia Papadopoulou-Mourkidou, "Occurrence of Pesticides in Rain of the Axios River Basin, Greece," *Environmental Science & Technology* 33, no. 14 (July 1999): 2363–68.

[7] "U.S. President's Cancer Panel Annual Report," 2010.

Rigorously Tested

*"All agricultural poisons are scientifically
tested to ensure safe use."*

One of the greatest pesticide myths is that all agricultural poisons are scientifically tested to ensure that they are used safely. According to the United States President's Cancer Panel (USPCP), this is simply not the case: "Only a few hundred of the more than 80,000 chemicals in use in the United States have been tested for safety."[1] The fact is that the overwhelming majority of chemicals used worldwide have not been subjected to testing. Given that according to the USPCP the majority of cancers are caused by environmental exposures, especially to chemicals, this oversight shows a serious level of neglect by regulatory authorities.

Pesticides have been subjected to more testing than most chemicals. However, where chemicals, including pesticides, have been subjected to testing, many leading scientists regard it as inadequate to determine whether they are safe for or harmful to humans. The USPCP report states: "Some scientists maintain that current toxicity testing and exposure limit-setting methods fail to accurately represent the nature of human exposure to potentially harmful chemicals."[2]

There are several key areas in particular in which many experts and scientists believe testing has not sufficiently established that the current use of pesticides and other chemicals is safe.

CHEMICAL COCKTAILS IN FOOD AND WATER

Regulatory authorities approve multiple pesticides for a crop on the basis that all of them can be used in normal production. Consequently a mixture of several different toxic chemical products is applied during the normal course of agricultural production for most foods, including combinations of herbicide products, insecticide products, fungicide products, and synthetic fertilizer compounds. A substantial percentage of foods thus have a cocktail of small amounts of these toxic chemicals that we absorb through food, drink, dust, and the air. According to the USPCP, "Only 23.1 percent of [food] samples had zero pesticide residues detected, 29.5 percent had one residue, and the remainder had two or more."[3] This means that about half the foods in the United States contain a mixture of chemical residues. Because people consume a variety of foods, with around 77 percent containing residues of different types of agricultural chemicals, most people's normal dietary habits include consuming a chemical concoction of which they are unaware.

A study by the U.S. Centers for Disease Control found a cocktail of toxic chemicals in the blood and urine of most Americans that they tested.[4] In 2009 the Environmental Working Group found up to 232 chemicals in the placental cord blood of newborns in the United States.[5] Many of these pollutants have been linked to serious health risks such as cancer and can persist for decades in the environment.

Regulatory authorities assume that because each of the active ingredients in individual commercial products is below the acceptable daily intake (ADI), the cocktail is thus also safe. They do not test these combinations of chemicals—the chemical cocktails that are ingested daily by billions of people—to ensure that they are safe.

Several scientific studies raise serious concerns. The emerging body of evidence demonstrates that many chemical cocktails can act synergistically, meaning that instead of one plus one equaling two, the extra effect of the mixtures can lead to one plus one equaling five or even higher in toxicity and damaging health effects.

The World Health Organization (WHO) and the United Nations Environment Programme (UNEP) published a comprehensive meta-analysis on endocrine- (hormone-) disrupting chemicals titled *State of the Science of Endocrine Disrupting Chemicals 2012*. Over sixty recognized international experts worked throughout 2012 to contribute to the meta-analysis to ensure that it was an up-to-date compilation of the current scientific knowledge on endocrine disruptors. This meta-study questioned the practice of testing single chemicals in isolation and ignoring the potential dangers posed by a cocktail of chemicals. "When the toxicity of chemicals is evaluated, their effects are usually considered in isolation, with assumptions of 'tolerable' exposures derived from data about one single chemical. These assumptions break down when exposure is to a large number of additional chemicals that also contribute to the effect in question."[6]

The WHO and UNEP study showed that an additive effect occurred when estradiol (a form of the female sex hormone, estrogen) was combined with other chemicals capable of mimicking estrogen. When each chemical was tested individually at low levels they did not produce any observable effect; however when they were combined they produced considerable adverse effects. According to the study,

> For a long time, the risks associated with these "xenestrogens" [artificial estrogens] have been dismissed, with the argument that their potency is too low to make an impact on the actions of estradiol. But it turned out that xenestrogens, combined in sufficient numbers and at concentrations that on their own do not elicit measureable effects, produced substantial estrogenic effects. . . . When mixed together with estradiol, the presence of these xenestrogens at low levels even led to a doubling of the effects of the hormone (Rajapakse, Silva & Kortenkamp, 2002)."[7]

A number of scientific studies detail the synergistic and/ or additive effects of chemical cocktails in which the cocktail causes health problems even though testing each of the chemicals individually deemed that they were safe. A study called "Endocrine, Immune and Behavioral Effects of Aldicarb (Carbamate), Atrazine (Triazine) and Nitrate (Fertilizer) Mixtures at Groundwater Concentrations" in the journal *Toxicology and Industrial Health* showed that combinations of low doses of commonly used agricultural chemicals can significantly affect health. In the experiments conducted by Porter et al. at the University of Wisconsin–Madison, mice were given drinking water containing combinations of pesticides, herbicides, and nitrate fertilizer at concentrations currently found in groundwater in the United States. The mice exhibited altered immune, endocrine (hormone), and nervous system functions. The effects were most noticeable when a single herbicide (atrazine) was combined with nitrate fertilizer.[8]

Atrazine is widely used in agricultural industries in conjunction with synthetic fertilizers that add nitrate to the soil. It is also one of the most persistent herbicides, measurable in corn, milk, beef, and many other foods in the United States. "The U.S. Geological Survey's [USGS] national monitoring study found atrazine in rivers and streams, as well as groundwater, in all thirty-six of the river basins that the agency studied. It is also often found in air and rain; USGS found that atrazine was detected in rain at nearly every location tested. Atrazine in air or rain can travel long distances from application sites. In lakes and groundwater, atrazine and its breakdown products are persistent, and can persist for decades."[9]

XENOESTROGENS are found in pesticides and insecticides like DDT, glyphosate, and endosulfan and have been linked to breast cancer and precocious puberty.

In Europe atrazine was found in most water courses and in a significant percentage of rain samples.[10] The European Union and Switzerland consequently banned it to prevent this widespread pollution, but it is still broadly used in many countries, and in some cases, as in the United States, is one of the most common herbicides.

The research by Porter et al. showed that the influence of pesticide, herbicide, and fertilizer mixtures on the endocrine system may also cause changes in the immune system and affect fetal brain development. Of particular concern was thyroid disruption in humans, which has multiple consequences including effects on brain development, level of irritability, sensitivity to stimuli, ability or motivation to learn, and an altered immune function.

A later experiment in 2002 by Cavieres et al. found that very low levels of a mixture of the common herbicides 2,4-D, Mecoprop, Dicamba, and inert ingredients caused a decrease in the number embryos and lives births in mice at all doses tested. Very significantly, the data showed that even low and very low doses caused these problems.[11]

Research conducted by Laetz et al. and published in *Environmental Health Perspectives* studied the combinations of common pesticides that were found in salmon habitats and found that these combinations could have synergistic effects. There was a greater degree of synergistic effects at higher doses. The scientists found that several combinations of organophosphate pesticides were lethal at concentrations that had been sublethal in single chemical trials. The researchers concluded that current risk assessments used by regulators underestimated the effects of these insecticides when they occurred in combinations.[12]

One of the most concerning studies, by Manikkam et al., found that exposure to a combination of small amounts of common insect repellents, plasticizers, and jet fuel residues during pregnancy can induce permanent changes in the germ line (the first cells that lead to the formation of sperm or egg production cells) of the fetus. The researchers found that these changes are inherited by future generations.[13]

A similar study investigated short-term exposure of pregnant female rats to a mixture of a fungicide, a pesticide mixture, a plastic mixture, dioxin, and a hydrocarbon at the time when the fetus was starting sex determination of the gonads. The researchers found that the next three generations had an increase in cysts, resembling human polycystic ovarian disease, and a decrease in the ovarian primordial follicle pool size, resembling primary ovarian insufficiency in humans.[14] The researchers also found that the exposure had changed the way certain genes operated and that this change was passed on to future generations, an effect caused by several different classes of chemicals. The scientists stated, "Epigenetic transgenerational inheritance of ovarian disease states was induced by all the different classes of environmental compounds, suggesting a role of environmental epigenetics in ovarian disease etiology."[15] Epigenetics is the study of environmental factors that cause changes in the way genes express their traits without any changes in the DNA of the genes.

A 2009 study by the Environmental Working Group found up to

232 CHEMICALS

in the placental cord blood of newborns in the United States. Many of these chemicals, such as mercury and polychlorinated biphenyls, are known to **HARM BRAIN DEVELOPMENT AND THE NERVOUS SYSTEM**.

These studies raised two new and very concerning issues, firstly the health effects that may occur when low-level residues of common pesticides are combined with minute levels of residues of the numerous other types of common chemicals that are found in the environment and in humans. This is an area that has been largely neglected by the research community and completely ignored by all regulatory authorities, but it is a major concern in the context of multiple U.S. studies. A study by the U.S. Centers for Disease Control found a cocktail of many toxic chemicals in the blood and urine of most Americans. As previously mentioned, a 2009 study by the Environmental Working Group found up to 232 chemicals in the placental cord blood of newborns in the United States.[16] Many of these chemicals, such as mercury and polychlorinated biphenyls, are known to harm brain development and the nervous system. These studies show the inaccuracy of the regulatory authorities' assumption that because each of these chemicals is present at a low level in commercial products they will cause no health issues. This assumption clearly has no basis in science. Regulatory authorities should be making their decisions and taking appropriate actions based on scientific evidence, not on data-free assumptions.

Secondly, the fact that the researchers found that these changes are inherited by future generations is a major issue in terms of the lasting and widespread health damage that is most likely being inflicted on human society. Regulatory authorities should be taking urgent action to prevent this rather than ignoring the danger.

THE COMBINATION OF THE PESTICIDES PRODUCED BY GMO PLANTS WITH HERBICIDES

Another area emerging as a concern is the combination of the pesticides produced by GMO plants (*Bacillus thurengiensis*, or Bt) with the herbicides and other pesticides used in crop production. The pesticide-producing GMO crops do not eliminate pesticide usage. They may reduce some types of pesticide usage, but some studies show an increased usage of pesticides with GMO plants, especially herbicide usage.[17]

A peer-reviewed, published study that researched the combination of the GMO-produced Bt toxin pesticides and Roundup found that they altered the normal life cycle of cells in human organs. The researchers concluded: "In these results, we argue that modified Bt toxins are not inert on nontarget human cells, and that they can present combined side effects with other residues of pesticides specific to GM plants."[18]

SINGLE-CHEMICAL TESTING IS INADEQUATE

The ever-increasing body of peer-reviewed science shows that the current methodology of only testing the active ingredient as a single agent and not testing common combinations is flawed and insufficient to determine the safety of chemical exposure in a real-world situation where humans are exposed to daily cocktails of chemicals.

The USPCP clearly states, "In addition, agents are tested singly rather than in combination. Single-agent toxicity testing and reliance on animal testing are inadequate to address the backlog of untested chemicals already in use and the plethora of new chemicals introduced every year."[19]

REGISTERED AGRICULTURAL PRODUCTS

The overwhelming majority of registered pesticide products used in agriculture as insecticides, herbicides, and fungicides are formulations of several chemicals. They are mixtures composed of one or more chemicals that are defined as the active ingredient(s) or active principle and are combined with other mostly toxic chemicals, such as solvents, adjuvants, and surfactants, that are defined as inerts.

The active ingredient is the primary chemical that acts as the pesticide. The other chemicals in the mixture are called inerts as they have a secondary role in the formulation. The name "inert" is misleading as most of these other compounds are chemically active in their functions in the pesticide formulations. They help to make the active ingredient work more effectively. According to the USPC report, many of these "inert" ingredients are toxic; however, they are not tested for their potential to cause health problems. "Pesticides (insecticides, herbicides, and fungicides) approved for use by the U.S. Environmental Protection Agency (EPA) contain nearly 900 active ingredients, many of which are toxic. Many of the solvents, fillers, and other chemicals listed as inert ingredients on pesticide labels also are toxic, but are not required to be tested for their potential to cause chronic diseases such as cancer."[20]

An example is Roundup and other glyphosate-based herbicide formulations. These pesticides are a mixture of glyphosate as the active ingredient and inerts such as ammonium sulfate, benzisothiazolone, glycerine, isobutane, isopropylamine, polyethoxylated alkylamines, polyethoxylated tallowamine POE-15, and sorbic acid.[21] Glyphosate barely works as an herbicide without the assistance of the inerts to boost its effectiveness.

Some studies show an increased usage of pesticides with

GMO PLANTS,

especially herbicide usage.

The active ingredient is the only chemical in the formulation that is tested for some of the known health problems caused by chemicals—such as cancer, organ damage, birth defects, and cell mutations—to determine a safe level for the acceptable daily intake (ADI) and the maximum residue limit (MRL). The complete pesticide formulation of the active ingredient and the "inerts" is not tested for health problems.

ACUTE TOXICITY AND LD_{50}

There are a limited number of registered products in which the whole formulation is tested for acute toxicity, or the amount of the product that is fatal to animals and humans. The most referenced value in acute toxicity tests is LD_{50}, which stands for lethal dose (LD) 50 percent or median lethal dose. This number represents the milligrams of the chemical per kilogram of body mass needed to kill 50 percent of the test animals. The lower the number, the more toxic the chemical because a smaller amount is needed to kill the animals. LD_{50} 100 milligrams per kilogram is more toxic than LD_{50} 400 milligrams per kilogram because only a quarter of the amount is needed to kill the same amount of animals.

LD_{50}s are widely used as the main reference when judging a substance's acute toxicity, or the adverse effects resulting from either a single exposure or multiple exposures in a short span of time. Adverse effects must occur within two weeks of the chemical being administered to be considered in acute toxicity. LD_{50}s are thus irrelevant in showing the longer-term toxic effects of a chemical or compound.[22] These are the toxicities that cause other health issues such as cancers, cell mutations, endocrine disruption, birth defects, organ and tissue damage, nervous system damage, behavior changes, and immune system damage.

Asbestos is a good example of how measuring only LD_{50} can be misleading about a chemical or compound's toxicity. Asbestos does not have an LD_{50} because it is not acutely toxic. It is not a poison in the traditional sense. It is technically possible to eat asbestos by the bucket load and not be poisoned. However, a minute speck of asbestos dust entering the lungs can result in three fatal diseases: asbestosis, lung cancer, and mesothelioma. As early as the 1920s and 1930s there were studies linking asbestos to health problems. Asbestos is a classic case of regulatory neglect and industries misrepresenting the dangers of their products.

The fact that asbestos is not toxic under the LD_{50} criteria was used by the asbestos industry and government regulators for decades to deny that it was a dangerous product, resulting in the widespread and irresponsible use of asbestos in houses, schools, offices, cars, boats, hairdryers, and numerous other applications. Most communities are sitting on ticking time bombs healthwise, with numerous people in many countries dying from asbestos-related illnesses. The huge costs of removing and disposing of it into toxic waste dumps fall on governments and communities rather than the companies that profited from mining and selling it.

It took decades of activism by concerned scientists, nongovernmental organizations, and consumers before regulatory authorities took action to ban asbestos. In the meantime many thousands of people died unnecessary, cruel deaths, and many thousands more are yet to die this way because of the twenty- to fifty-year latency period for asbestos-related diseases.

Consumers and industries alike should consider the tragedies of asbestos a warning about regulatory neglect of published science.

SCIENTIFICALLY UNSOUND METHODOLOGY
Many scientists and researchers consider it scientifically unsound to test just one component of a mixture and assume that the whole combination of chemicals in a formulation will respond in the same

way. Despite the limited testing, there are some studies that compare the differences in toxicity between the active ingredient and the registered formulated product. Glyphosate-based herbicides are amongst the most studied for these effects.

There are numerous studies that show that Roundup is more toxic than its active ingredient, glyphosate. These studies link the pesticide to a range of health problems such as cancer, placental cell damage, miscarriages, stillbirths, endocrine disruption, and damage to various organs such as the kidney and liver.[23]

Research by scientists in France has shown that one of the "inert" adjuvants in Roundup, the polyethoxylated tallowamine POE-15, is considerably more toxic to human cells than the "active" ingredient glyphosate. The researchers found that at one and three parts per million (ppm), doses that are considered to be normal environmental and occupational exposures, POE-15 enters human cells and causes them to die. This is a different mode of action from glyphosate, which is known to promote endocrine- (hormone-) disrupting effects after entering cells. The scientists stated, "Altogether, these results challenge the establishment of guidance values such as the acceptable daily intake of glyphosate, when these are mostly based on a long term in vivo test of glyphosate alone. Since pesticides are always used with adjuvants that could change their toxicity, the necessity to assess their whole formulations as mixtures becomes obvious. This challenges the concept of active principle of pesticides for non-target species."[24]

In the only study where nine formulated pesticides were tested on human cells at levels well below agricultural dilutions, the research scientists found that eight of the nine formulations were several hundred times more toxic than their respective active ingredients. The researchers stated, "Adjuvants in pesticides are generally declared as inerts, and for this reason they are not tested in long-term regulatory experiments. It is thus very surprising that they amplify up to 1000 times the toxicity of their AP [active ingredient] in 100% of the cases where they are indicated to be present by the manufacturer."[25]

Fungicides were the most toxic to human cells, even at concentrations three hundred to six hundred times lower than agricultural dilutions, followed by herbicides and then insecticides. Roundup was the most toxic of the herbicides and insecticides they tested. The scientists concluded "Our results challenge the relevance of the Acceptable Daily Intake for pesticides because this norm is calculated from the toxicity of the active principle alone."[26]

None of the formulated registered pesticide products are tested for the numerous types of health problems that can be caused by chemicals. ADIs and MRLs are not set for any these formulated products. They are only set for the "active" ingredient.

> **FUNGICIDES** were the most toxic to human cells, even at concentrations three hundred to six hundred times lower than agricultural dilutions.

It should be of great concern to everyone that the vast majority of the nearly 1,400 registered pesticide and veterinary products used in the United States, around 7,000 used in Australia, and the countless thousands used worldwide for the production of food have had no testing for numerous health and environmental problems linked to the exposure to cocktails of chemicals.[27]

All countries share this practice, other than the European Union, which has started a process of assessing over 143,000 chemicals and chemical formulations.[28]

Given the body of scientific data linking the additive and synergistic effects of chemical mixtures to numerous adverse health effects, serious concerns need to be raised as to why regulators allow these formulated mixtures to be used on the assumption that they are safe. There are no credible scientific data to determine a safety level for the residues of the actual registered pesticide products used in food production and found in food until whole formulations are tested.

THE SPECIAL NEEDS OF THE
DEVELOPING FETUS AND NEWBORN

The USPCP and many scientific researchers have expressed concern that the current toxicology testing methodologies are grossly inadequate for children.

The USPCP report stated, "They [children] are at special risk due to their smaller body mass and rapid physical development, both of which magnify their vulnerability to known or suspected carcinogens, including radiation."[29]

This is a critically important issue given that, according to the USPCP, "Approximately 40 chemicals classified by the International Agency for Research on Cancer (IARC) as known, probable, or possible human carcinogens, are used in EPA-registered pesticides now on the market."[30]

The main food regulator in Australia and New Zealand, Food Standards Australia and New Zealand (FSANZ), acknowledged that children had the highest levels of dietary exposure to pesticides when they published the 20th Australian Total Diet Survey due to their size and weight ratios in relation to the amount of residues they receive from food. "In general, the dietary exposure to pesticide residues was highest for the toddler age group. This is due to the high food consumption relative to body weight."[31] FSANZ, along with most regulators, are not concerned about this because pesticide residues in food are usually below the maximum residue limits. However the USPCP and other scientific researchers have pointed out that the current testing protocols are based on testing mature animals and ignore the specific physiological differences between mature animals and the fetus, newborns, and developing young, including humans.

According to the USPCP, "Chemicals typically are administered when laboratory animals are in their adolescence, a methodology that fails to assess the impact of in utero, childhood, and lifelong exposures."[32]

This is a critical issue as there is a large body of published science showing that the fetus and the newborn are continuously being exposed to numerous chemicals. The USPCP stated, "Some of these chemicals are found in maternal blood, placental tissue, and breast milk samples from pregnant women and mothers who recently gave birth. These findings indicate that chemical contaminants are being passed on to the next generation, both prenatally and during breastfeeding."[33]

The U.S. President's Cancer Panel not only expressed concern on the level of these chemical contaminants, they also pointed out that this issue is being ignored by regulators due to the critical lack of knowledge and researchers. "Numerous environmental contaminants can cross the placental barrier; to a disturbing extent, babies are born 'pre-polluted.' Children also can be harmed by genetic or other damage resulting from environmental exposures sustained by the mother (and in some cases, the father). There is a critical lack of knowledge and appreciation of environmental threats to children's health and a severe shortage of researchers and clinicians trained in children's environmental health."[34]

Dr. Theo Colborn, one of the world's acknowledged leading experts on endocrine-disrupting chemicals and coauthor of *Our Stolen Future*, published a peer-reviewed study in the scientific journal *Environmental Health Perspectives* that examined these issues. The study reviewed many of the scientific papers and showed the widespread extent to which children and the unborn are exposed to numerous pesticides. Multiple pesticide residues have been found in semen, ovarian follicular fluid, amniotic fluid, maternal blood, placental and umbilical cord blood, breast milk, meconium of newborns, and in the urine of children. She writes, "It is fairly safe to say that every child conceived today in the Northern hemisphere is exposed to pesticides from conception throughout gestation and lactation regardless of where it is born."[35]

The information from these numerous scientific studies shows that current regulatory systems around the world have failed to protect unborn and growing children from exposure to a massive cocktail of toxic pesticides. This has many serious implications, including an increase in a range of serious health issues in children and as adults later in life.

CHILDREN'S CANCER RATES ARE INCREASING

A number of studies show the link between chemical exposure, particularly exposure to pesticides, and the increase of cancer in children. The USPCP report states, "Cancer incidence in U.S. children under 20 years of age has increased. . . . Leukemia rates are consistently elevated among children who grow up on farms, among children whose parents used pesticides in the home or garden, and among children of pesticide applicators."[36]

NERVOUS SYSTEM DAMAGE

Many pesticides work as nerve poisons. These include organophosphates, synthetic pyrethroids, neonicotinoids, and carbamates. Organophosphates were first developed by German chemists in the 1930s looking to use them as pesticides. They were further developed by the Nazis as nerve gases for warfare in World War II, although it is doubtful they were ever used then. One of the best known organophosphate nerve gases is sarin, which was used to kill thirteen people in the Tokyo subway attack by the Aum Shinrikyo religious sect on March 20, 1995, and injure nearly a thousand. Saddam Hussein used a range of organophosphate nerve gases such as sarin and VX gas during the Iran-Iraq War to kill Iranian soldiers and on his own citizens, killing thousands of Kurds. The United Nations (UN) has stated that sarin gas was used by the Syrian government in the Damascus suburb of Ghouta in August 2013, killing an estimated 281 to 1,729 rebel fighters and civilians. The production and stockpiling of chemical weapons, including sarin, was banned in 1993 by the UN Chemical Weapons Convention. Organophosphates started to become a major class of pesticides after World War II with the commercialization of numerous types such as malathion, parathion, diazinon, chlorpyrifos, azamethiphos, dichlorvos, phosmet, fenitrothion, fenthion, dimethoate, omethoate, tetrachlorvinphos, etc.

Organophosphates react with and destroy a key nervous system enzyme called acetylcholinesterase. This enzyme is responsible for degrading acetylcholine, one of the neurotransmitter chemicals that fire nerve signals like bullets fired from a gun. Acetylcholine is found mostly in the muscle nerves and in the brain. Without acetylcho-

> **LEUKEMIA RATES** are consistently elevated among children who grow up on farms, among children whose parents used pesticides in the home or garden, and among children of pesticide applicators.

linesterase to "turn off" acetylcholine, the nerves continue to fire signals, causing a range of symptoms such as intense headaches, nausea, vomiting, muscular paralysis, convulsions, and bronchial constriction. High levels of exposure can cause death from asphyxiation. Low levels of exposure are usually associated with "flu-like" symptoms—headaches, low energy, depression, and a general feeling of being unwell.

Standard toxicology usually regards the reactive degradation of acetylcholinesterase as the only way organophosphates poison animals and posits that they do not damage other metabolic pathways or body organs. Acetylcholinesterase levels will generally recover after low-level exposures, so it is assumed that no permanent damage results from such contact.

There are studies showing that organophosphate pesticides damage other tissues, including the myelin (the protective covering of nerve cells), and key nerves such as the optic nerve, causing permanent damage to eyesight including blindness. Other studies show genetic damage to the cell chromosomes. This is usually regarded as a sign of a precancerous condition.[37]

Dr. Colborn reviewed numerous published papers on one of the most common organophosphates, chlorpyrifos (CPF). These papers detailed an amazing litany of diverse mechanisms in the way CPF affected many tissues and the nervous system, raising serious questions about the safety of CPF, other organophosphates, and all pesticides. These effects included damage to several areas of the brain and disruption of the development of the nervous system in the fetus and newborn that resulted in a range of behavioral problems later in life. She states, "Most astounding is the fact that a large part of CPF's toxicity is not the result of cholinesterase inhibition, but of other newly discovered mechanisms that alter the development and function of a number of regions of the brain and CNS [central nervous system]."[38]

DEVELOPMENTAL NEUROTOXICITY

Scientific research shows that many pesticides affect the normal development of the nervous system in fetuses and children. The brain is the largest collection of nerve cells, and there are several scientific studies showing that when the fetus and the newborn are exposed to minute amounts of these pesticides, even below the current limits set by regulatory authorities, brain function can be significantly altered.

Qiao et al. of the Department of Pharmacology and Cancer Biology at the Duke University Medical Center found that the developing fetus and the newborn are particularly vulnerable to pesticides in amounts lower than the levels currently permitted by most regulatory authorities around the world. Their studies showed that the fetus and the newborn possess lower concentrations of the protective serum proteins than adults.[39] A major consequence is developmental neurotoxicity, in which the poison damages the developing nervous system.[40] The scientists stated, "These results indicate that chlorpyrifos and other organophosphates such as diazinon have immediate, direct effects on neural cell replication.... In light of the protective effect of serum proteins, the fact that the fetus and newborn possess lower concentrations of these proteins suggests that greater neurotoxic effects may occur at blood levels of chlorpyrifos that are nontoxic to adults."[41] Contact with chemicals at levels below the regulatory limits can thus harm the fetus and breastfeeding children even if the mother shows no side effects from the contact.

One of the most concerning studies on this matter was published in 1998 by Guillette et al. in the peer-reviewed scientific journal *Environmental Health Perspectives*. The researchers compared the drawing abilities of four- and five-year-old Yaqui children in the Sonora region of Mexico. The study compared two groups of children that shared similar diets, genetic backgrounds, and cultural backgrounds. One group lived in the valley and was exposed to the drift of pesticides from the surrounding farms, and the other lived in the foothills where they were not so exposed. Both groups of children were asked to draw pictures of people. The children from the foothills drew pictures consistent with children their age. The children exposed to pesticides could not draw an image or could barely draw an image that represented a person. Most of their drawings resembled the scribbles of much younger children, indicating that pesticide exposure had severely compromised the development of their brain functions.[42]

BRAIN ABNORMALITIES AND IQ REDUCTIONS IN CHILDREN

Concerns raised by Guillette's study about the development of brain function were validated by four later studies that showed that prenatal exposure to organophosphate insecticides (OPs) adversely affects the neurological development of children. The studies were conducted by researchers at the Columbia University Center for Children's Environmental Health, the University of California, Berkeley, and the Mount Sinai School of Medicine. Each study was conducted independently, but they all came up with very similar results: fetal exposure to small amounts of OPs will reduce the IQs of children.

A study of farm worker families in California has shown that by age three and a half, children born to mothers exposed to OP insecticides have lessened attention spans and are more vulnerable to attention deficit hyperactivity disorder (ADHD). Male children were more likely to be impacted. According the Centers for Disease Control and Prevention, Atlanta, Georgia, "The average annual increase in the percentage of children with all diagnoses of ADHD (with and without LD [learning difficulties]) was 3% from 1997 through 2006. No significant average annual change was found in the percentage of children with all diagnoses of LD (with and without ADHD)."[43] While the overall trend for all types of learning difficulties is stable, the trend for ADHD is steadily increasing.

Parents should have considerable concern that the Columbia University study found no evidence of a lower-limit threshold of

PESTICIDE EFFECTS ON CHILDREN

Some examples from Dr. Elizabeth Guillette's study on the effects of pesticides on children. The children from the valley had been exposed to pesticides far more frequently than the children from the foothills, and the difference in the drawings between the two groups is clear. The exercise revealed a correlation between pesticide exposure and impaired development and motor skills.

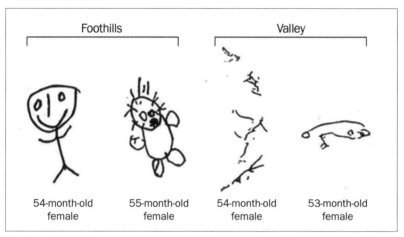

Representative drawings of a person by four-year-old Yaqui children from the valley and foothills of Sonora, Mexico.

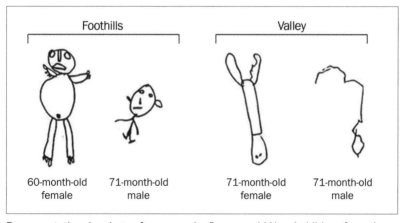

Representative drawings of a person by five-year-old Yaqui children from the valley and foothills of Sonora, Mexico.

Source: Guillette et al., "An Anthropological Approach to the Evaluation of Preschool Children Exposed to Pesticides in Mexico," Environmental Health Perspectives *106, no. 6 (June 1998): 347–53*

exposure to organophosphates in the observed adverse impact on intelligence. This means that even very low levels of exposure could lead to reductions in a child's intelligence.[44]

The study by Rauh et al., published in the journal *Proceedings of the National Academy of Sciences of the United States of America*, has confirmed the findings of the previous studies and shown a large range of brain abnormalities present in children exposed to chlorpyrifos in utero through normal, nonoccupational uses. Exposure to CPF in the womb, even at normal levels, resulted in "significant abnormalities in morphological measures of the cerebral surface associated with higher prenatal CPF exposure" in a sample of forty children between five and eleven years old.[45]

The researchers stated that the current regulatory safety limits and testing methodologies are inadequate for determining safe exposure levels for children.

> Current safety limits are set according to levels needed to achieve inhibition of plasma cholinesterase, a surrogate for inhibition of acetylcholinesterase in the brain, long assumed to be the common mechanism by which organophosphates induce neurodevelopmental deficits. However, pathogenic mechanisms other than cholinesterase inhibition are almost certainly contributing to the deleterious effects of early exposure to organophosphates (21, 37), including the observed brain abnormalities and their accompanying cognitive deficits.[46]

Measuring the levels of cholinesterase in blood is the accepted method used to establish the level of exposure to organophosphates. If the levels are considered "normal" then it is assumed that there is no damage to the developing nervous system and the brain. The researchers state that the current assumption that the degradation of acetylcholinetserase is the only way that organophosphates affect the nervous system is incorrect. The abnormalities that the researchers found in the developing brains of children along with the cognitive deficits show that organophosphates have other mechanisms that cause damage.

The researchers recommended that the current limits set by regulatory authorities be revised based on these data. "Human exposure limits based on the detection of cholinesterase inhibition may therefore be insufficient to protect brain development in exposed children."[47]

The United States had banned all uses of chlorpyrifos on food and restricted it to non-food uses prior to these studies. Despite this effort, the exposure levels continue to cause neurodevelopmental problems in U.S. children. The U.S. EPA is now reviewing all uses of chlorpyrifos.

This study has even greater implications for the many countries that allow the use of chlorpyrifos in food crops, especially in the horticulture sector. It is one of the pesticides that are regularly detected in residue surveys of food. Children of these countries almost certainly have had a higher level of exposure to chlorpyrifos.

Another study published in *Environmental Health Perspectives* looked at a range of chemicals, including organophosphate pesticides, that were implicated in lowering the IQs of zero- to five-year-old U.S. children. It found that the reduction in IQ was substantial. The study concluded that "when population impact is considered, the contributions of chemicals to FSIQ [full-scale IQ points] loss in children are substantial, in some cases exceeding those of other recognized risk factors for neurodevelopmental impairment in children. The primary reason for this is the relative ubiquity of exposure."[48]

> Studies have indicated that exposure to chemicals has a greater harmful impact on
>
> # CHILDREN'S INTELLIGENCE
>
> because these chemicals are now so ever-present in the environment.

In a study published in the medical journal *The Lancet Neurology*, scientists from the University of Southern Denmark, the Harvard School of Public Health in Boston, and the Icahn School of Medicine at Mount Sinai, New York, expressed concern that the majority of commercially used chemicals, including pesticides, have not been tested for developmental neurotoxicity. The researchers noted that neurodevelopmental disabilities, including autism, attention-deficit hyperactivity disorder, dyslexia, and other cognitive impairments, affected millions of children worldwide. They stated, "To control the pandemic of developmental neurotoxicity, we propose a global prevention strategy. Untested chemicals should not be presumed to be safe to brain development, and chemicals in existing use and all new chemicals must therefore be tested for developmental neurotoxicity."[49]

A large body of published, peer-reviewed scientific research shows that pesticide exposure in children is linked to:

- Lower IQs
- ADHD
- Autism spectrum disorders
- Lack of physical coordination
- Loss of temper/anger management issues
- Bipolar/schizophrenia spectrum of illnesses
- Depression

The previous studies show that the current methods of determining the MRLs and ADIs for organophosphate and other pesticides are clearly out of date and need to be immediately revised based on the warnings of current, peer-reviewed science. There is an urgent need to investigate the many other biochemical pathways other than acetylcholinesterase that organophosphate and other neurotoxic pesticides can affect.

DAMAGE PASSED ON TO FUTURE GENERATIONS
Some of the most concerning studies show that pesticide damage can be passed on to the next generation. Not only are the offspring

born with damage to the nervous system, the reproductive system, and other organs, the great-grandchildren can be as well.[50]

Researchers in a 2012 study found that pregnant rats and mice exposed to the fungicide vinclozolin during the period when the fetus was developing reproductive organs developed genetic changes in the genes that were passed onto future generations. The researchers stated, "Transient exposure of the F0 generation* gestating female during gonadal sex determination promoted transgenerational adult onset disease in F3 generation male and female mice, including spermatogenic cell defects, testicular abnormalities, prostate abnormalities, kidney abnormalities and polycystic ovarian disease. Pathology analysis demonstrated 75% of the vinclozolin lineage animals developed disease with 34% having two or more different disease states."[51]

Another study showed that when pregnant rats were exposed to a combination of permethrin, a common insecticide, and DEET (N,N-diethyl-meta-toluamide), the most common insect repellent, significant damage occurred in subsequent generations, including the great-grandchildren. The researchers found that "Gestating F0 generation female rats were exposed during fetal gonadal sex determination and the incidence of disease evaluated in F1 and F3 generations. There were significant increases in the incidence of total diseases in animals from pesticide lineage F1 and F3 generation animals. Pubertal abnormalities, testis disease, and ovarian disease (primordial follicle loss and polycystic ovarian disease) were increased in F3 generation animals."[52]

The significant issue with these two studies is that small exposures to pesticides at critical times in the development of the fetus can cause multiple diseases that are passed on to future generations. It means that pregnant women eating food with minute levels of pesticides could be inadvertently exposing their children, grandchildren and great-grandchildren to permanent damage to their reproductive systems and other organs.

The scientists found that 363 genes had been altered by the pesticides: "Analysis of the pesticide lineage F3 generation sperm epigenome identified 363 differential DNA methylation regions

---

* F0 is the first generation, F1 their children, F2 their grandchildren, and so on.

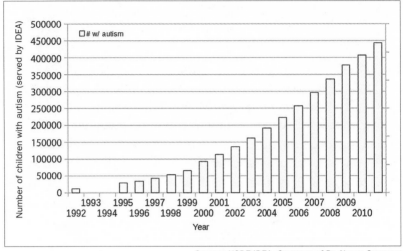

Source: USDE:IDEA, Courtesy of Dr. Nancy Swanson

(DMR) termed epimutations [changes to the genes caused by environmental factors—in this case by pesticides]. Observations demonstrate that a pesticide mixture (permethrin and DEET) can promote epigenetic transgenerational inheritance of adult onset disease and potential sperm epigenetic biomarkers for ancestral environmental exposures."[53]

Genes play a major role in the development of hormone, metabolic, reproductive, nervous, and other body systems, and in the development of organs, limbs, the brain, and other body parts. When genes are altered, this means that the development of the systems and organs that are dependent on the genes are altered, leading to a range of diseases and other problems later in life.

This study is particularly distressing because DEET is the most common repellent used for mosquitoes and other insects. It is widely used on children and pregnant women.

PUBLISHED, PEER-REVIEWED SCIENCE IS SUBSTANTIAL AND COMPELLING

The body of published, peer-reviewed science showing the wide range of problems caused by pesticides to the fetus and newborn is substantial and compelling. The current testing methodologies use adolescent through to adult animals. This means that they will not detect the adverse health issues that are specific to the unborn and small children.

Despite the fact that many professional experts in this area such as the USPCP, the WHO, and UNEP have been calling for specific toxicological studies that are relevant to the fetus and growing children to determine if the current MRLs and ADIs of pesticides are safe for them, regulatory authorities largely continue to ignore this dangerous oversight. Until these specific tests are done, regulatory authorities are using data-free assumptions that the current pesticides used in food, households, playgrounds, schools, and in the general environment are safe for our unborn and growing children. Regulation should not be based on assumptions but should use independently published, peer-reviewed scientific evidence to prove whether these toxins are safe. Given that our children are our future and most of them are exposed to multiple chemicals, time will prove the regulatory committees' unwillingness to act against pesticides more serious than their decades of inaction over asbestos.

NOTES

[1] "U.S. President's Cancer Panel Annual Report," 2010.

[2] Ibid.

[3] Ibid.

[4] Margo Higgins, "Toxins Are in Most Americans' Blood, Study Finds," *Environmental News Network*, March 26, 2001; "U.S. President's Cancer Panel Annual Report," 2010.

[5] "U.S. President's Cancer Panel Annual Report," 2010.

[6] Bergman et al., *State of the Science of Endocrine Disrupting Chemicals 2012.*

[7] Ibid.

[8] Warren P. Porter, James W. Jaeger, and Ian H. Carlson, "Endocrine, Immune and Behavioral Effects of Aldicarb (Carbamate), Atrazine (Triazine) and Nitrate (Fertilizer) Mixtures at Groundwater Concentrations," *Toxicology and Industrial Health* 15 (January 1999): 133–50.

[9] Caroline Cox, "Atrazine: Environmental Contamination and Ecological Effects," Northwest Coalition against Pesticides, Eugene, Oregon, *Journal of Pesticide Reform* 21, no. 3 (Fall 2001): 12.

[10] Charizopoulos and Papadopoulou-Mourkidou, "Occurrence of Pesticides in Rain of the Axios River Basin, Greece."

[11] María Fernanda Cavieres, James Jaeger, and Warren Porter, "Developmental Toxicity of a Commercial Herbicide Mixture in Mice: I. Effects on Embryo Implantation and Litter Size," *Environmental Health Perspectives* 110, no. 11 (November 2002): 1081–85.

[12] Cathy A. Laetz et al., "The Synergistic Toxicity of Pesticide Mixtures: Implications for Risk Assessment and the Conservation of Endangered Pacific Salmon," *Environmental Health Perspectives* 117, no. 3 (March 2009): 348–53.

[13] Mohan Manikkam et al., "Transgenerational Actions of Environmental Compounds on Reproductive Disease and Identification of Epigenetic Biomarkers of Ancestral Exposures," *PLoS ONE* 7, no. 2 (February 2012).

[14] Eric Nilsson et al., "Environmentally Induced Epigenetic Transgenerational Inheritance of Ovarian Disease." *PLoS ONE* 7, no. 5 (May 2012): e36129. doi:10.1371/journal.pone.00361.

[15] Ibid.

[16] "U.S. President's Cancer Panel Annual Report," 2010.

[17] Charles Benbrook, "Impacts of Genetically Engineered Crops on Pesticide Use in the U.S.—The First Sixteen Years," *Environmental Sciences Europe* 24, no. 24 (September 2012): doi:10.1186/2190-4715-24-24; Jack A. Heinemann et al., "Sustainability and Innovation in Staple Crop Production in the US Midwest," *International Journal of Agricultural Sustainability*, published online June 14, 2013, http://www.tandfonline.com/doi/full/10.10 80/14735903.2013.806408#.Ut766dLnaU.

[18] Robin Mesnage et al., "Cytotoxicity on Human Cells of Cry1Ab and Cry1Ac Bt Insecticidal Toxins Alone or with a Glyphosate-Based Herbicide," *Journal of Applied Toxicology* 33, no. 7 (July 2013): 695–99. Originally published online February 2012.

[19] "U.S. President's Cancer Panel Annual Report," 2010.

[20] Ibid.

[21] Caroline Cox, "Glyphosate (Roundup)," Northwest Coalition against Pesticides, Eugene, Oregon, *Journal of Pesticide Reform* 24, no. 4 (Winter 2004).

[22] Short, *Quick Poison, Slow Poison*.

[23] Cox, "Glyphosate (Roundup)"; Sophi Richard, Safa Moslemi, Herbert Sipahutar, Nora Benachour, and Gilles-Éric Serálini, "Differential Effects of Glyphosate and Roundup on Human Placental Cells and Aromatase," *Environmental Health Perspectives* 113, no. 6 (June 2005): 716–20. Published online February 25, 2005, http://www.ncbi.nlm.nih.gov/pmc/articles/PMC1257596/; Robin Mesnage, Benoît Bernay, and Gilles-Éric Séralini, "Ethoxylated Adjuvants of Glyphosate-Based Herbicides are Active Principles of Human Cell Toxicity," *Toxicology* 313, nos. 2–3 (November 2013): 122–28. Published online September 21, 2012, http://dx.doi.org/10.1016/j.tox.2012.09.006.

[24] Mesnage et al., "Ethoxylated Adjuvants of Glyphosate-Based Herbicides."

[25] Robin Mesnage et al., "Major Pesticides Are More Toxic to Human Cells than Their Declared Active Principles," *BioMed Research International* (December 2013), http://www.hindawi.com/journals/bmri/aip/179691/.

[26] Ibid.

[27] "U.S. President's Cancer Panel Annual Report," 2010; Australian Pesticides and Veterinary Medicines Authority (APVMA), "About the APVMA: Factsheet," September 2008, http://www.apvma.gov.au/publications/fact_sheets/docs/about_apvma.pdf.

[28] European Commission, "What is REACH?," January 24, 2014, http://ec.europa.eu/environment/chemicals/reach/reach_en.htm.

[29] "U.S. President's Cancer Panel Annual Report," 2010.

[30] Ibid.

[31] Food Standards Australia and New Zealand, "20th Australian Total Diet Survey," 2002, available online at http://www.foodstandards.gov.au/publications/Pages/20thaustraliantotaldietsurveyjanuary2003/20thaustraliantotaldietsurveyfullreport/Default.aspx.

[32] "U.S. President's Cancer Panel Annual Report," 2010.

[33] Ibid.

[34] Ibid.

[35] Theo Colborn, "A Case for Revisiting the Safety of Pesticides: A Closer Look at Neurodevelopment," *Environmental Health Perspectives* 114, no. 1 (January 2006): 10–17.

[36] "U.S. President's Cancer Panel Annual Report," 2010.

[37] Cox, "Glyphosate (Roundup)."

[38] Colborn, "A Case for Revisiting the Safety of Pesticides."

[39] Dan Qiao, Frederic Seidler, and Theodore Slotkin, "Developmental Neurotoxicity of Chlorpyrifos Modeled In Vitro: Comparative Effects of Metabolites and Other Cholinesterase Inhibitors on DNA Synthesis in PC12 and C6 Cells," *Environmental Health Perspectives* 109, no. 9 (September 2001): 909–13.

[40] Justin Aldridge et al., "Serotonergic Systems Targeted by Developmental Exposure to Chlorpyrifos: Effects during Different Critical Periods," *Environmental Health Perspectives* 111, no. 14 (November 2003): 1736–43; Gennady A. Buznikov et al., "An Invertebrate Model of the Developmental Neurotoxicity of Insecticides: Effects of Chlorpyrifos and Dieldrin in Sea Urchin Embryos and Larvae," *Environmental Health Perspectives* 109, no. 7 (July 2001): 651–61; Gertrudis Cabello et al., "A Rat Mammary Tumor Model Induced by the Organophosphorous Pesticides Parathion and Malathion, Possibly through Acetylcholinesterase Inhibition," *Environmental Health Perspectives* 109, no. 5 (May 2001): 471–79.

[41] Qiao, Seidler, and Slotkin, "Developmental Neurotoxicity of Chlorpyrifos Modeled In Vitro."

[42] Elizabeth A. Guillette et al., "An Anthropological Approach to the Evaluation of Preschool Children Exposed to Pesticides in Mexico," *Environmental Health Perspectives* 106, no. 6 (June 1998): 347–53.

[43] Patricia N. Pastor and Cynthia A. Reuben, "Diagnosed Attention Deficit Hyperactivity Disorder and Learning Disability: United States, 2004–2006," National Center for Health Statistics, *Vital and Health Statistics* 10, no. 237 (July 2008).

[44] Virginia Rauh et al., "Brain Anomalies in Children Exposed Prenatally to a Common Organophosphate Pesticide," *Proceedings of the National Academy of Sciences of the United States of America* 109, no. 20 (May 2012), www.pnas.org/cgi/doi/10.1073/pnas.1203396109; Maryse F. Bouchard et al., "Prenatal Exposure to Organophosphate Pesticides and IQ in 7-Year-Old Children," *Environmental Health Perspectives* 119, no. 8 (August 2011): 1189–95, published online April 21, 2011, http://www.ncbi.nlm.nih.gov/pmc/articles/PMC3237357/; Stephanie M. Engel et al., "Prenatal Exposure to Organophosphates, Paraoxonase 1, and Cognitive Development in Children," *Environmental Health Perspectives* 119 (2011): 1182–88, published online April 21, 2011, http://ehp.niehs.nih.gov/1003183/.

[45] Rauh et al., "Brain Anomalies in Children Exposed Prenatally."

[46] Ibid.

[47] Ibid.

[48] David C. Bellinger, "A Strategy for Comparing the Contributions of Environmental Chemicals and Other Risk Factors to Neurodevelopment of Children," Environmental Health Perspectives 120, no. 4 (2012): 501–7, http://dx.doi.org/10.1289/ehp.1104170.

[49] Philippe Grandjean and Philip J. Landrigan, "Neurobehavioural Effects of Developmental Toxicity," *The Lancet Neurology* 13, no. 3 (March 2014): 330–38.

[50] Manikkam et al., "Transgenerational Actions of Environmental Compounds on Reproductive Disease; Carlos Guerrero-Bosagna et al., "Epigenetic Transgenerational Inheritance of Vinclozolin Induced Mouse Adult Onset Disease and Associated Sperm Epigenome Biomarkers," *Reproductive Toxicology* 34, no. 4 (December 2012): 694–707; Mohan Manikkam et al., "Pesticide and Insect Repellent Mixture Permethrin and DEET Induces Epigenetic Transgenerational Inheritance of Disease and Sperm Epimutations," *Journal of Reproductive Toxicology* 34, no. 4 (December 2012): 708–19.

[51] Guerrero-Bosagna et al., "Epigenetic Transgenerational Inheritance."

[52] Manikkam et al., "Pesticide and Insect Repellent Mixture Permethrin and DEET."

[53] Ibid.

Very Small Amount

*"The residues are too low to
cause any problems."*

Most developed nations have pesticide residue programs in which food is periodically tested for pesticide residues exceeding the maximum residue limits (MRLs) and the average daily intake (ADI). This testing shows that the pesticide residues in food are generally below the MRLs and ADIs. It is on the basis of this testing that the regulatory authorities state that the food is safe, as the residues are too low to cause any problems.

There are, however, serious concerns about the way these MRLs and ADIs are set, several of which were explained in the previous chapter.

The growing body of published science on endocrine disruption, in which very small amounts of some types of synthetic chemicals can act like hormones and disrupt our hormone systems, is another area where many scientists are concerned that the current methods of assessing the toxicity of synthetic chemicals in our diet and environment are out of date and grossly inadequate in determining safe levels of exposure.

Dose Responses

Dose responses measure how a particular amount of a substance affects us. See the graphs in this chapter for a visual example of these response curves.

Monotonic dose response: Response decreases as dosage decreases
Non-linear dose response: Response does not change in direct proportion to dosage amount
Non-monotonic dose response: Response begins to decrease as dosage decreases, but then begins to increase as dosage continues to decrease.

The 2013 meta-study by the WHO and UNEP clearly states that there are many gaps in the current testing methods used to determine the safety of chemicals, including that the current tests are not able to screen chemicals for hormone-disrupting effects. "Perhaps most importantly, the exposure periods do not cover critical developmental windows of increased susceptibility now known to exist."[1]

The current model of toxicology (science of poisons) works on the notion that the lower the dose, the less the effect of the poison. When animal testing shows that a certain dose level of poison causes no observable adverse effects (NOAEL), this dose becomes the basis that is used to determine the average daily intake (ADI). The ADI is usually determined by lowering the permitted amount by a factor of a hundred or a thousand times lower. The regulatory authorities then claim that any residue levels below the ADI are too low to cause health problems. This model is based on the assumption that the toxic effect decreases with lower dosage in a steady linear progression until the compound is no longer toxic. It comes from the maxim of Paracelsus, the sixteenth-century physician and father of toxicology, who stated that "All things are poison and nothing is without poison; only the dose makes a thing not a poison." This has been condensed to: "The dose makes the poison."

In the 1990s this four-hundred-year-old concept was proven incorrect for many chemicals through the endocrine disruption evidence presented in two books: *Our Stolen Future* and *The Feminization of Nature*. The peer-reviewed science summarized in these books showed that many chemicals, including agricultural chemicals, were mimicking hormones like estrogen.[2]

There are numerous exceptions to the assumption of a steady linear decrease in toxicity; one of the most profound is the evidence of non-monotonic responses in many chemicals when they start to act as hormones at very low levels.

The lowest doses of some chemicals can be more toxic instead of the least toxic. The current regulatory methodology of determining the ADI by lowering the threshold level of exposure is therefore problematic. This threshold is determined on the assumption that all chemicals including pesticides continue to decrease in toxicity in a linear model. Very little actual testing has been done at these levels to verify that this assumption is correct when setting the ADI.

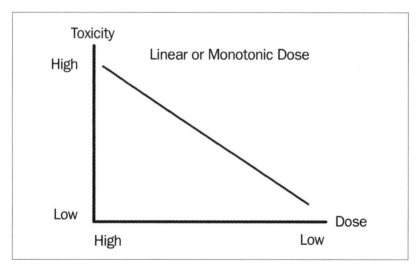

The graph above shows the standard model for toxicity of a steady linear decrease in the toxic effects in relation to the dose. The highest levels of toxic effects are at the highest doses. The toxic effect steadily decreases as the dose decreases.

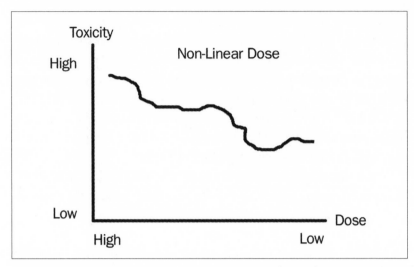

The graph above shows a non-linear dose. Instead of a predictable steady decrease of the toxicity in relation to the dose, as is found in the linear model, there can be an irregular decrease with areas where the toxicity stays around the same level, even though the dose is steadily decreasing.

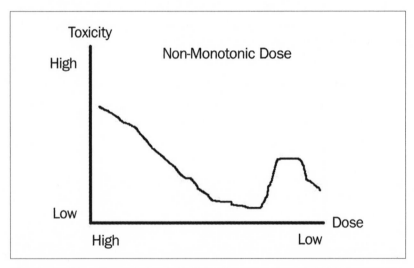

The graph above shows a non-monotonic response. This is where the toxicity may steadily decrease as the dose decreases; however instead of continuing to decrease as the dose decreases like in the linear model, there is a point or points at which the toxicity can increase as the dose decreases.

A significant meta review was published in March 2012 by several of the world's leading expert scientists in this field in the peer-reviewed journal *Endocrine Reviews*. Vandenberg et al. showed that there were hundreds of published studies documenting non-monotonic and non-linear doses where chemicals were more toxic at low and often at the lowest doses.[3] The scientists stated:

> We provide a detailed discussion of the mechanisms responsible for generating these phenomena, plus hundreds of examples from the cell culture, animal, and epidemiology literature. We illustrate that nonmonotonic responses and low-dose effects are remarkably common in studies of natural hormones and EDCs [endocrine-disrupting chemicals]. Whether low doses of EDCs influence certain human disorders is no longer conjecture, because epidemiological studies show that environmental exposures to EDCs are associated with human diseases and disabilities. We conclude that when nonmonotonic dose-response curves occur, the effects of low doses cannot be predicted by the effects observed at high doses.[4]

Evidence of non-monotonic dose response is of critical importance as it means that the ADIs and MRLs set for chemicals, including pesticides, have no actual scientific testing to determine that they do not adversely affect health outcomes by interfering with hormones. There are numerous studies showing that chemicals, including many pesticides, can be even more toxic at lower thresholds than the ADI, even though there were no observable adverse impacts at the higher dose levels that were used to set the NOAEL. Without testing, there is no way to know if the extrapolated assumptions used to set the ADI and consequently the MRL are correct and the recommendations safe.

No low threshold level should be assumed by extrapolating data from experiments done at higher doses. "Experimental data indicate that EDCs and hormones do not have NOAELs or threshold doses, and therefore no dose can ever be considered safe."[5] The

WHO and UNEP meta-study on endocrine disruption clearly makes this point: "Endocrine disruptors produce nonlinear dose responses both in vitro [using components of an organism] and in vivo [living organisms in their normal state]; these non linear dose responses can be quite complex and often include non-monotonic dose responses. They can be due to a variety of mechanisms; because endogenous hormone levels fluctuate, no threshold can

Evidence of non-monotonic dose response is of critical importance as it means that the **ADIs AND MRLs SET FOR CHEMICALS,** including pesticides, have no actual scientific testing to determine that **they do not adversely affect health outcomes** by interfering with hormones.

be assumed."[6] All the current ADIs and MRLs for pesticides need to undergo testing for endocrine disruption at the threshold levels that have been by set regulators to determine if they are truly safe.

THE ENDOCRINE SYSTEM

The endocrine system is based on numerous hormones that regulate the normal functioning and cycles of all living species including humans. This includes reproductive hormones such as estrogen, progesterone, and testosterone; growth hormones such as somatrophin; metabolic hormones such as dopamine and thyroid-stimulating hormones; circadian-rhythm hormones like melatonin; and pancreatic hormones like insulin. There are numerous hormones or hormone-related compounds in all living species, and these need to be at the correct levels to ensure the good health and well-being of plants, animals, and humans. If the levels of any hormones are too high or too low, they can cause a wide variety of diseases. All living species have inbuilt regulatory systems to moderate hormone levels to ensure that they are in balance and in a state called homoestasis. Good health requires that homeostasis is maintained.

In the 1940s scientists began to notice that some pesticides produced hormonal changes in test animals. By the 1980s there were many studies showing that numerous chemicals, including pesticides, were causing significant hormonal changes in living species.

As an example, researchers have now found that there are many chemicals that act like reproductive hormones, such as estrogen, the main female hormone. There are other chemicals that can work against hormones, such as chemicals that interfere with testosterone, the male hormone. These are known as anti-androgens. These chemicals can cause a range of reproductive and other problems in potentially all species, including humans. There are numerous studies showing that very low doses of many common pesticides and numerous other chemicals disrupt the endocrine system by acting as or affecting hormones.

According to the WHO and UNEP study, these endocrine-disrupting chemicals are linked to a range of reproductive and other problems in humans. The study found that up to 40 percent of young men in some countries have low semen quality. It also indicated an increase in genital malformations in baby boys such as undescended testes and penile malformations. These chemicals are also linked to an increase in adverse pregnancy outcomes such as preterm birth and low birth weight.

The increase in neurobehavioral disorders in children is associated with thyroid disruption. The age of breast development in girls is decreasing, and this is considered a risk factor for developing breast cancer later in life. Breast, endometrial, ovarian, prostate, testicular, and thyroid cancers are increasing. These are endocrine system–related cancers.

Obesity and type 2 diabetes levels have increased at a rapid rate around the world. There are 1.5 billion people who are overweight or obese, which is significantly more than the 850 million people who are undernourished. Between 1980 and 2008 the number of people with type 2 diabetes increased from 153 million to 347 million.[7]

There is a range of factors—such as diet, age, genetic makeup, sexually transmitted diseases, and exercise—that contribute to the increases in reproductive problems; however, they only explain part

ANNUAL INCIDENCE OF DIABETES (AGE ADJUSTED)

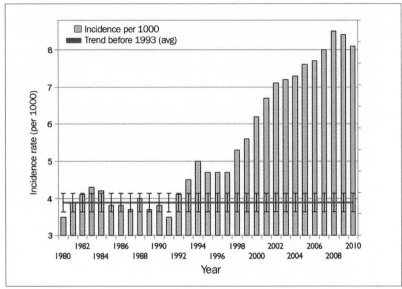

Sources: CDC, Courtesy of Dr. Nancy Swanson

AGE ADJUSTED DEATHS DUE TO OBESITY (ICD E66 & 278)

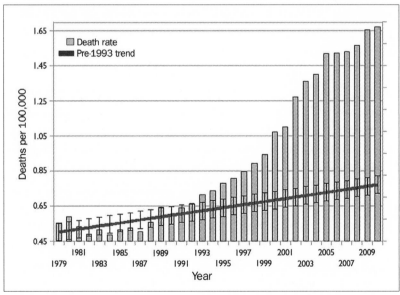

Sources: CDC, Courtesy of Dr. Nancy Swanson

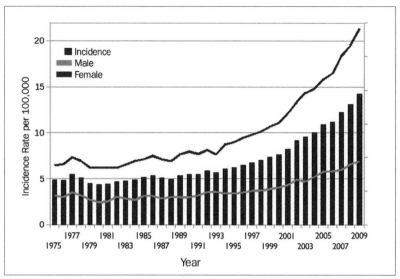

AGE ADJUSTED THYROID CANCER INCIDENCE RATE

Source: NCI:SEER, Courtesy of Dr. Nancy Swanson

of the increase. Numerous scientific studies show that EDCs cause these types of reproductive health problems in a wide range of animals, both in the wild and in laboratory research. Many scientists believe that ubiquitous exposure to EDCs is a significant reason for the increase in these widespread health problems. The WHO and UNEP meta-analysis stated:

> Moreover, effects of chemicals seen in exposed wildlife and in laboratory animals, similar to those seen in human populations and in DES-exposed individuals [diethylstilbestrol—a synthetic estrogen-mimicking chemical], have caused the scientific community to consider whether endocrine disruptors could also cause an increasing variety of reproductive health problems in women, including altered mammary gland development, irregular or longer fertility cycles, and accelerated puberty (Crain et al., 2008; Diamanti-Kandarakis et al., 2009; Woodruff et al., 2008). These changes indicate a higher risk of later health problems such as breast cancer, changes in lactation, or reduced fertility.[8]

The latest data from the International Agency for Research on Cancer (IARC), which is the specialized cancer agency of the WHO, showed an alarming rise in breast cancer rates in women. According to the IARC, "In 2012, 1.7 million women were diagnosed with breast cancer and there were 6.3 million women alive who had been diagnosed with breast cancer in the previous five years. Since the 2008 estimates, breast cancer incidence has increased by more than 20%, while mortality has increased by 14%. Breast cancer is also the most common cause of cancer death among women (522,000 deaths in 2012) and the most frequently diagnosed cancer among women in 140 of 184 countries worldwide."[9]

Many researchers have found that although some of these synthetic chemicals were considered relatively safe in parts per million in the standard toxicology tests, at doses of parts per billion or parts per trillion (more than a thousand times lower) they acted like hormones. This was because at the very low levels they could attach to hormone receptors, whereas at higher levels they were "ignored" by the hormone receptors. When these chemicals attach to hormone receptors, they send signals to the endocrine system whereby they either act as the hormone, as an antagonist to the hormone, or block the normal working of the hormone. This disrupts the normal signaling functions of the hormone (endocrine) system and thus the name "endocrine disruptor." A good analogy for understanding how this process works is a lock on a door that can only be opened by a specifically shaped key. A metal rod larger than the key if placed against the lock cannot open the door because it is too large to fit into the lock. A smaller wire bent into the correct shape, although different in shape from the key, can work as a lock pick. The wire can also just sit in the lock and prevent the normal key from opening it.

Most receptors are shaped so that only the specific hormone can fit into them and act as the key to "unlock" them and send the signal that will activate the specific response in a cell, tissue, organ, or gene.

When chemicals are given in higher doses they, like the metal bar, cannot fit into the receptor, so they are ignored. Many chemi-

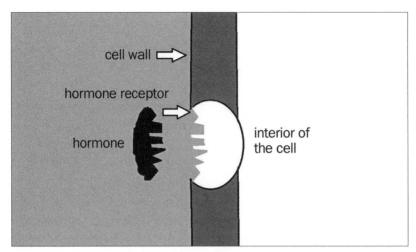

Each hormone receptor has a specific shape that allows the hormone to fit into it like a key in a lock. The specific shape of the receptor prevents other hormones or chemicals from fitting into to it.

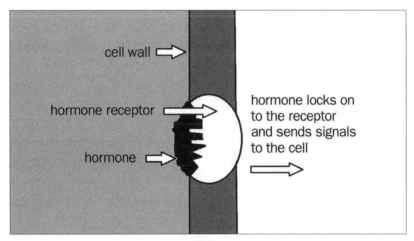

When the hormone fits into the receptor, it sends signals to the cell that regulates it. Hormones or chemicals of the wrong shape cannot fit into the receptor and therefore cannot send signals to the cell.

cals at significantly lower doses are analogous to the thin wire lock pick. They can fit into the receptor and send out hormone signals or just block the normal hormone signals. Because they are not the actual hormone, these signals are artificial, and so they disrupt the normal hormone function.

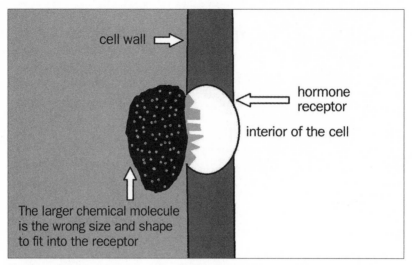

cell wall ⟹

hormone receptor

interior of the cell

The larger chemical molecule is the wrong size and shape to fit into the receptor

The larger-sized molecule cannot fit into the receptor and consequently no hormone signals are sent into the cell, so the cell is not disrupted with false instructions.

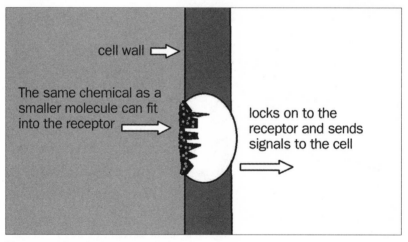

cell wall ⟹

The same chemical as a smaller molecule can fit into the receptor ⟹

locks on to the receptor and sends signals to the cell ⟹

The same chemical as a smaller molecule can fit into the receptor and either send false signals to the cell or block the normal hormone signals, thereby disrupting the normal functioning of the cell.

THE EFFECT ON THE UNBORN AND GROWING CHILDREN

A large body of research shows that fetuses, newborns, and growing children are the most vulnerable group when exposed to low levels of EDCs. The WHO and UNEP meta-study found that there are "particularly vulnerable periods during fetal and postnatal life when EDCs alone, or in mixtures, have strong and often irreversible effects on developing organs, whereas exposure of adults causes lesser or no effects. Consequently, there is now a growing probability that maternal, fetal and childhood exposure to chemical pollutants play a larger role in the etiology of many endocrine diseases and disorders of the thyroid, immune, digestive, cardiovascular, reproductive and metabolic systems (including childhood obesity and diabetes) than previously thought possible."[10]

The fetus is most vulnerable during the times when genes are turned on to develop specific organs. Small amounts of hormones signal genes to start developing various body parts and systems such as the reproductive tract, the nervous system, the brain, immune system, hormone systems, limbs, etc. Small disruptions in these hormone signals can significantly alter the way these body parts and systems will develop, and these altered effects will not correct themselves.

> This does not diminish their [EDCs] importance [in adults], but contrasts with their effects in the fetus and neonate where a hormone can have permanent effects in triggering early developmental events such as cell proliferation or differentiation. Hormones acting during embryonic development can cause some structures to develop (e.g. male reproductive tract) or cause others to diminish (e.g. some sex-related brain regions). Once hormone action has taken place, at these critical times during development, the changes produced will last a lifetime.[11]

The actions of hormones on the development of endocrine and physiological systems in fetuses are considered to be programming events. They set how these systems will function in adults. "Thus, small perturbations in estrogen action during fetal development can change the reproductive axis in adulthood and diminish fertility (Mahoney & Padmanabhan, 2010). It is now clear that fetal programming events can predispose the adult to a number of chronic diseases (Janesick & Blumberg, 2011; Hanson & Gluckman, 2011); thus, endocrine disease prevention should begin with maternal and fetal health."[12]

This is why the unborn are far more vulnerable to endocrine-disrupting chemicals than adults and why testing must be undertaken to determine the effects on the fetus and not just on adult and adolescent animals as is currently done.

There are new testing methods using human cell cultures that can reduce and in many cases end the need for live animal testing, which is now being criticized due to humane and ethical issues. Because these cell lines come from humans, in many cases they can give more accurate results as well as sets of data that are not available from live animal testing.

EXAMPLES OF A SMALL NUMBER OF THE HUNDREDS OF SCIENTIFIC STUDIES

The following are examples of a small number of the hundreds of scientific studies published on the effects of some of the common endocrine-disrupting pesticides and chemicals.

A study published in the peer-reviewed scientific journal *Food and Chemical Toxicology* in 2013 found that glyphosate at residue levels commonly found in people induced human breast cancer cells to multiply. The scientists found that these low levels of glyphosate caused a five- to thirteenfold increase in the multiplication of estrogen-sensitive breast cancer cells; however they had no effect on non-estrogen-sensitive breast cancer cells. The majority of human breast cancers are sensitive to estrogen. This means that estrogen and compounds that act as estrogen cause these types of cancers to grow. The researchers state, "Glyphosate exerted proliferative effects only in human hormone-dependent breast cancer, T47D cells, but not

> Glyphosate at residue levels commonly found in people induced human **BREAST CANCER** cells to multiply five- to thirteenfold.

in hormone independent breast cancer, MDA-MB231 cells, at 10-12 to 10-6 M in estrogen withdrawal condition."[13]

One of the major concerns is that this activity occurred at residue levels of glyphosate that are commonly found in the urine of most people and below the current safety levels set by regulators. "Concentrations of glyphosate tested in this study that exhibited estrogenic activity and interfered with normal estrogen signaling were relevant to the range of concentrations that has been reported in environmental conditions and exposed human. These results indicated that low and environmentally relevant concentrations of glyphosate possessed estrogenic activity."[14]

The scientists found that when these small levels of glyphosate in herbicides were combined with the normal levels of genistein, a phytoestrogen found in soybeans, it increased the multiplication of the breast cancer cells. The researchers concluded, "This study implied that the additive effect of glyphosate and genistein in postmenopausal woman may induce cancer cell growth."[15]

This additive effect is a great concern considering the vast increase in the planting of glyphosate-resistant GMO soybean varieties, which are now being widely used in soybean products such as soy milk, tofu, soy sauce, miso, etc. The scientists cited consuming these soybean products as a possible cause of breast cancer. "Furthermore, this study demonstrated the additive estrogenic effects of glyphosate and genistein which implied that the use of glyphosate-contaminated soybean products as dietary supplements may pose a risk of breast cancer because of their potential additive estrogenicity."[16]

Two peer-reviewed studies conducted by Hayes et al. showed that levels of atrazine, down to a level a thousand times lower than what's currently permitted in our food and in the environment, caused severe reproductive deformities in frogs.[17] Sara Storrs and Joseph Kiesecker of Pennsylvania State University confirmed Hayes's research. They exposed tadpoles of four frog species to atrazine and found that "Survival was significantly lower for all animals exposed to 3 ppb [parts per billion] compared with either 30 or 100 ppb. . . . These survival patterns highlight the importance of investigating the impacts of contaminants with realistic exposures and at various developmental stages."[18]

A study by Newbold et al. published in *Birth Defects Research* found that exposing pregnant mice to one part per billion of the synthetic estrogen mimic diethylstilbestrol (DES) can lead to off-spring becoming severely obese in adulthood. Exposing the mice to 100 ppb DES resulted in the offspring being scrawny and underweight as adults.[19]

A scientific study published by Howdeshell et al. in the peer-reviewed journal *Nature* found that feeding pregnant mice 2.4 parts per billion of bisphenol A (BPA, an estrogenic plasticizing chemical used to make many plastic products including baby bottles and water bottles) on days eleven to seventeen of pregnancy resulted in the majority of the female offspring reaching sexual maturity earlier than the untreated females. The treated female offspring were also heavier than untreated females at sexual maturity. This is more than a thousand times lower than the fifty parts per million set by the U.S. EPA as safe.[20]

The study by Cavieres et al., published in 2002 in *Environmental Health Perspectives*, on the mixture of the common herbicides 2,4-D, mecoprop, dicamba, and inert ingredients showed that the greatest decrease in the number embryos and lives births in mice was at the lowest dosage level tested. The great concern is that the dose was one-seventh of the U.S. EPA level set for the drinking water.[21]

Skakkebæk et al. published a study in the journal *Human Reproduction* showing that a number of estrogenic and anti-androgenic chemicals, including common pesticides, are linked to abnormal changes in the development of the testes in the fetus. They found that these endocrine-disrupting chemicals were linked to the ever-increasing rate of genital urinary tract problems, deformities, and diseases such as undescended testes, low sperm counts, hypospadias, intersex, and testicular cancers.[22]

THE IMPLICATIONS

The substantial body of published scientific research on endocrine disruption has several far-reaching implications. Firstly, the current methods of toxicology testing used to permit chemical residues in our food and water are based on the assumption that all chemicals lose their toxicity as their levels decrease to the point where they are nontoxic. This is clearly not correct for many chemicals. The USPCP report states, "Some scientists maintain that current toxicity testing and exposure limit-setting methods fail to accurately represent the nature of human exposure to potentially harmful chemicals. Current toxicity testing relies heavily on animal studies that utilize doses substantially higher than those likely to be encountered by humans."[23] This report clearly shows the need to test chemicals at the levels found in food, the environment, and most importantly in the human body.

> Scientific testing has shown that the pesticides atrazine, chlorpyrifos, glyphosate, maneb, parathion, and vinclozolin are **ENDOCRINE DISRUPTORS** at very low doses.

Secondly, numerous published, peer-reviewed studies link the endocrine disruption caused by common commercially available chemicals to many of the health problems that are increasing in our society. These include the rise in obesity, type 2 diabetes, ADHD, depression, cancers of the sexual tissues and endocrine system, and genital-urinary tract malformations, as well as lowered fertility rates and sperm counts.

Thirdly, the results of these studies show that the current regulatory systems cannot guarantee the safe use of chemicals. The vast majority of the synthetic chemicals used in our food supply and in our environment, including pesticides, have not been tested for endocrine disruption. The USPCP report stated that the current testing methodologies fail to detect harmful effects that may come from very low doses: "These data—and the exposure limits extrapolated from them—fail to take into account harmful effects that may occur only at very low doses."[24]

The WHO and UNEP meta-study raised the same issue about the current testing methodologies: "Close to 800 chemicals are known or suspected to be capable of interfering with hormone receptors, hormone synthesis or hormone conversion. However, only a small fraction of these chemicals have been investigated in tests capable of identifying overt endocrine effects in intact organisms."[25] The study expresses great concern over the fact that most of the thousands of synthetic chemicals have not been tested at all. The study authors expressed further concerns that the lack of testing means there is no credible scientific data that can validate that the current use of these chemicals is safe. "This lack of data introduces significant uncertainties about the true extent of risks from chemicals that potentially could disrupt the endocrine system."[26]

> ## CLOSE TO 800 CHEMICALS
> are known or suspected to be capable of interfering with hormone receptors, hormone synthesis or hormone conversion. However, only a small fraction of these chemicals have been investigated in tests.

The meta-study by Vandenberg et al. stated that there is a need for changes in the methodologies used to test chemicals. Regulators need to take into account the possibility of non-monotonic dose responses when testing for safety rather than just assuming that the toxicity of all chemicals reduces as the dosage is lowered. Neglecting this possibility when testing commercially available chemicals poses serious hazards to human and environmental health. "Thus, fundamental changes in chemical testing and safety determination are needed to protect human health."[27] The WHO and UNEP meta-study also stated the need to change current methodologies so that tests can be developed for endocrine disruption.

Until thorough scientific testing is done on these chemicals and their effects on the endocrine system, regulatory authorities have no scientific basis or evidence supporting the assumption that exposure to chemical residues is safe at recommended levels. In the light of the hundreds of studies showing non-monotonic dose curves in endocrine-disrupting chemicals, setting low thresholds for ADIs and MRLs by extrapolating data from testing animals at higher levels of exposure on the assumption that the toxicity will decrease in a steady linear model is clearly not evidence-based science. It is decision making based on data-free assumptions.

NOTES

[1] Bergman et al., *State of the Science of Endocrine Disrupting Chemicals 2012*.

[2] Colborn, Dumanoski, and Myers, *Our Stolen Future*; Cadbury, *Feminization of Nature*.

[3] Laura N.Vandenberg et al., "Hormones and Endocrine-Disrupting Chemicals: Low-Dose Effects and Nonmonotonic Dose Responses," *Endocrine Reviews* 33, no. 3 (June 2012): 378–455.

[4] Ibid.

[5] Ibid.

[6] Bergman et al., *State of the Science of Endocrine Disrupting Chemicals 2012*.

[7] Ibid.

[8] Ibid.

[9] "Latest World Cancer Statistics, Global Cancer Burden Rises to 14.1 Million New Cases in 2012: Marked Increase in Breast Cancers Must Be Addressed," International Agency for Research on Cancer and the World Health Organization, press release, December 12, 2013.

[10] Ibid.

[11] Ibid.

[12] Ibid.

[13] Siriporn Thongprakaisang et al., "Glyphosate Induces Human Breast Cancer Cells Growth via Estrogen Receptors," *Food and Chemical Toxicology* 59 (September 2013): 129–36, http://dx.doi.org/10.1016/j.fct.2013.05.057.

[14] Ibid.

[15] Ibid.

[16] Ibid.

[17] Tyrone B. Hayes et al., "Hermaphroditic, Demasculinized Frogs after Exposure to the Herbicide Atrazine at Low Ecologically Relevant Doses," *Proceedings of the National Academy of Sciences* 99, no. 8 (April 2002): 5476–80; Tyrone B. Hayes et al., "Atrazine-Induced Hermaphroditism at 0.1 ppb in American Leopard Frogs (*Rana pipiens*): Laboratory and Field Evidence," *Environmental Health Perspectives* 111, no. 4 (April 2003): 569–75.

[18] Sara I. Storrs and Joseph M. Kiesecker, "Survivorship Patterns of Larval Amphibians Exposed to Low Concentrations of Atrazine," *Environmental Health Perspectives* 112, no. 10 (July 2004): 1054–57.

[19] Retha R. Newbold et al., "Developmental Exposure to Estrogenic Compounds and Obesity," *Birth Defects Research Part A: Clinical and Molecular Teratology* 73, no. 7 (2005): 478–480.

[20] Kembra Howdeshell et al., "Environmental Toxins: Exposure to Bisphenol A Advances Puberty," *Nature* 401 (October 1999): 762–64.

[21] Cavieres, "Developmental Toxicity of a Commercial Herbicide Mixture in Mice."

[22] N. E. Skakkebæk, E. Rajpert-De Meyts, and K. M. Main, "Testicular Dysgenesis Syndrome: An Increasingly Common Developmental Disorder with Environmental Aspects," *Human Reproduction* 16, no. 5 (2001): 972–78.

[23] "U.S. President's Cancer Panel Annual Report," 2010.

[24] Ibid.

[25] Bergman et al., eds., *State of the Science of Endocrine Disrupting Chemicals 2012*.

[26] Ibid.

[27] Vandenberg et al., "Hormones and Endocrine-Disrupting Chemicals."

Breakdown

"Modern pesticides rapidly biodegrade."

One of the major pesticide legends is the belief that most modern agricultural chemicals rapidly biodegrade and leave few if any residues. We are misled into believing that they break down and do not persist in our food like older chemicals such as DDT.

The following is a claim by the main food regulator in Australia and New Zealand, FSANZ, and is typical of the claims by many nations' regulators. They state, "Organophosphorous pesticides, carbamate pesticides are mostly biodegradable, and therefore do not concentrate in the food chain. Synthetic pyrethroids . . . are generally biodegradable and therefore tend not to persist in the environment."[1] These types of statements give the false impression that few agricultural pesticides persist in our food and environment. In fact, most agricultural and veterinary chemicals leave residues in food. That is the reason why tolerances for maximum residue limits (MRLs) and the acceptable daily intake (ADI) are set for these poisons.

The data presented in the United States President's Cancer Panel 2010 report indicating that only 23.1 percent of food samples had zero pesticide residues is reasonably consistent with the data from testing in most countries. This means that the overwhelming majority of foods contain pesticide residues.

Many of the current chemicals, including some of the synthetic pyrethroids, organophosphates, carbamates, and herbicides such as atrazine, are as residual as the mostly banned older chemicals such as the organochlorine group that includes dieldrin, DDT, chlordane, heptachlor, lindane, and aldrin.

METABOLITES OF PESTICIDES

One of the biggest myths is the assumption that once a chemical degrades it disappears and is harmless. Most agricultural poisons leave residues of breakdown products or daughter chemicals when they degrade.[2] These breakdown products of chemicals are also called metabolites. Where there is research, it shows that many of the metabolites from agricultural poisons cause health and reproductive problems.

A substantial number of agricultural pesticides—such as organophosphates like diazinon, malathion, chlorpyrifos, and dimethoate—become even more toxic when they break down. These

Oxons

Studies have shown that many pesticides used in agriculture, such as diazinon, malathion, chlorpyrifos, and dimethoate, become even more dangerous to the environment as they break down into metabolites called oxons. Oxons result when a chemical bond between phosphorus and sulfur is replaced by a bond between phosphorous and oxygen as the pesticide breaks down in the environment. Oxons can cause significant damage to animals' nervous systems.

metabolites are known as oxons. Scientists at the Cooperative Wildlife Research Laboratory at Southern Illinois University and the Western Ecology Research Center of the U.S. Geological Survey in Point Reyes, California, found that the oxons can be up to one hundred times more toxic than the original pesticide.

> In this study the oxon derivatives of chlorpyrifos, malathion, and diazinon were significantly more toxic than their respective parental forms. Chloroxon killed all of *R. boylii* tadpoles and was at least 100 times more [toxic] than the lowest concentration of chlorpyrifos which resulted in no mortality. Maloxon was nearly 100 times more toxic than malathion and diazoxon was approximately 10 times more toxic than its parent. This is consistent with other studies that have compared parent and oxon forms.[3]

Dimethoate is a good example. Dimethoate is a systemic pesticide because it is absorbed into all the tissues of the plant, including the edible portions such as all the flesh of fruits, stems, tubers, and leaves.

Contrary to popular belief, because systemic poisons are absorbed into the flesh—and consequently every part of the plant is toxic—washing or peeling the surface of the food only removes a small percentage of the poisons on the surface. It will not remove the bulk of poison, which is inside the food.

Dimethoate is widely used as a fruit fly treatment because it is so residual that even after two weeks any maggots that hatch from eggs inside the fruit will be killed by the poison residues in the edible portion of the flesh. Dimethoate breaks down to an even more toxic metabolite called omethoate. Omethoate is also used as a pesticide and consequently, unlike the vast majority of metabolites, it has been researched and has an LD_{50}. According to the WHO, omethoate has an LD_{50} of 50 milligrams per kilogram, whereas dimethoate has an LD_{50} of 150 milligrams per kilogram. This means that as the dimethoate decays within the treated food, it becomes 300 percent more toxic as omethoate. Under the WHO classifica-

> Contrary to popular belief, washing or peeling the surface of the food **ONLY REMOVES A SMALL PERCENTAGE OF THE PESTICIDES** on the surface. It will not remove the bulk of poison, which is inside the food.

tion of hazards it goes from being a moderately hazardous to a highly hazardous pesticide. Several countries have withdrawn or are in the process of withdrawing omethoate from use as a pesticide due to its high toxicity and its persistence. Other countries are still debating whether to ban dimethoate. All food that is treated with dimethoate will end up with residues of the more toxic and persistent omethoate as well as a number of other toxic metabolites that are generated as the dimethoate breaks down.

In her article "A Case for Revisiting the Safety of Pesticides," Dr. Theo Colborn gives the example of research into paraoxon, the main metabolite of parathion, showing that it is very toxic and causes a range of negative health effects. "Chronic paraoxon exposure (0.1, 0.15, or 0.2 mg/kg subcutaneously) during a stage of rapid cholinergic brain development from PND8 to PND20 [various stages of prenatal development] in male Wistar rats led to reduced dendritic spine density in the hippocampus without obvious toxic cholinergic signs in any of the animals (Santos et al. 2004). Some animals in the two highest dose groups died in the early days of the study. All doses caused retarded perinatal growth, and brain cholinesterase activity was reduced 60% by PND21."[4]

Glyphosate is another example. It breaks down into the more persistent aminomethylphosphonic acid (AMPA) that has been linked to liver disease.[5]

A scientific study published in the journal *Annals of Allergy, Asthma & Immunology* found that exposure to dichlorophenols was linked to an increase in food allergies. Dichlorophenols are metabolites of chlorinated pesticides such as 2,4-D, dichlorvos, and chlorpyrifos, and they are found in chlorinated drinking water. The researchers concluded that "High urine levels of dichlorophenols are associated with the presence of sensitization to foods in a US population. Excessive use of dichlorophenols may contribute to the increasing incidence of food allergies in westernized societies."[6]

IMPURITIES IN PESTICIDES

Pesticide testing is done with pure, laboratory-grade active ingredients and not with actual ingredients from the mass-manufacturing process. Manufacturing processes can result in the creation of a number of by-products, many of which can be toxic. "Other industrial chemicals or processes have hazardous by-products or metabolites. Numerous chemicals used in manufacturing remain in or on the product as residues, while others are integral components of the products themselves."[7] These by-products are largely ignored by regulatory authorities based on the assumption that, because they are at such low levels, they are safe. However where there has been testing, some of these impurities have been found to be highly toxic.

Dioxins, or more correctly polychlorinated dibenzodioxins (PCDDs), are examples of some of the most common impurities. PCDDs are commonly called dioxins because their primary molecules have dioxin skeletal rings. There are potentially hundreds of dioxins, most of which have had limited testing. Dioxins are one of the major groups of metabolites that result from chemical processes that use chlorine. These can include chlorine bleaching fibers for paper or textiles, the wood preservative pentachlorophenol, herbicides such as 2,4-D and pesticides such chlorpyrifos.

Dioxins can be generated by burning or heating substances that contain chlorine, as in municipal and hospital wastes and crop residues that have been treated with pesticides containing chlorine. Some of the major emitters are sugar mills that burn the crop residues that have been treated with chlorinated herbicides and pesticides such as 2,4-D and chlorpyrifos as the energy source to boil the sugar cane juice in the first stage of sugar production.

Some forms of dioxins are among the most toxic chemicals known to science and can cause a wide variety of illnesses, especially cancers and birth defects. Chlorine is a common ingredient in many pesticides due to its toxicity and its residual persistence.

Dioxins are also endocrine disrupters, and according to the study by the WHO and UNEP they cause sex ratio imbalances in humans and wildlife, resulting in fewer males. "EDC-related sex ratio imbalances, resulting in fewer male offspring in humans, do exist (e.g., in relation to dioxin and 1,2-dibromo-3-chloropropane), although the underlying mechanisms are unknown. The effects of dioxin on sex ratio are now corroborated by results obtained in the mouse model."[8]

Agent Orange, an herbicide that was widely used to destroy the highly biodiverse rainforests in Vietnam and Laos during the Vietnam War, was the best known of the chemicals contaminated with dioxins. Agent Orange was a combination of two herbicides: 2,4-D and 2,4,5-T. The manufacturing process of 2,4,5-T resulted in very high levels of dioxins, particularly 2,3,7,8-tetrachlorodibenzo-p-dioxin (TCDD). This was the reason it was banned. However 2,4-D continues to be widely used despite being contaminated with TCDD, one of the most toxic dioxins. It also contains other dioxins. These dioxins are present as impurities from the manufacturing processes; however, they can also be formed as metabolites as the 2,4-D decays.

Dioxins are very persistent in the environment, so consequently Vietnam still has extremely high levels of the environmental contamination resulting in birth defects, immune diseases, cancers, and many other problems more than forty years after the widespread use of Agent Orange was stopped.

Dioxins are pervasive throughout the global environment and are found in the tissues of most living species, especially in species at the top of the food chain, such as humans, as they bioaccumulate. In some cases they can come from natural causes, such as active volcanoes and forest fires; however the bulk of dioxins are by-products of the chemical industry.

Some of the most infamous cases are Love Canal, New York; Times Beach, Missouri; and the massive release from an industrial accident in Seveso, Italy. The attempted assassination of President

Viktor Yushchenko of Ukraine by poisoning with dioxins in 2004 resulted in permanent health problems. Due to a lack of research, however, the full extent of the contribution of the numerous chlorinated pesticides to the widespread global contamination by dioxins in the environment and the tissues of most living species has not been determined.

The greatest concern, according to the U.S. National Institute of Environmental Health Sciences (NIEHS), is that most people are exposed to dioxins from food, "in particular animal products, contaminated by these chemicals. Dioxins are absorbed and stored in fat tissue and, therefore, accumulate in the food chain. More than 90 percent of human exposure is through food."[9] Examples include dioxins being found in mozzarella cheese in Italy and in pork in Ireland in 2008 and in animal feed in Germany in 2010.

LACK OF COMPLETE TESTING

To the knowledge of this author, no country in the world has tested food for every pesticide used, and most only test for a "representative sample" of commonly used pesticides. For example, very few if any national residue monitoring programs test for glyphosate due to the difficulty posed by testing for it, despite the fact that it is the most commonly used herbicide. There is virtually no testing to detect the residues of the metabolites and by-products of agricultural poisons in our food and water. The 23 percent of food in the United States that was found with no residues could still be toxic for two reasons.

Firstly, the 23 percent of food with no residue is largely meaningless if the testing does not include 100 percent of pesticides used in food production. How can the researchers claim that the food is free of residues if they have not tested for every possible residue? Secondly, just because the testing didn't find the parent chemical does not mean that it is free of the toxic residues of the metabolites or the toxic by-products that can result from the manufacturing of pesticides. All it means is that there has been no testing for them.

MORE RESEARCH NEEDED TO
DETERMINE METABOLITE SAFETY

Very little research has been done to determine safe intake levels for the metabolites or the by-products of agricultural poisons. Consequently there are virtually no safety levels to determine the average daily intake (ADI) for the numerous toxic metabolites and by products that contaminate our food.

The toxicity and health effects of pesticide metabolites and impurities are mostly ignored on an assumption that they are safe. The regulation of pesticides is supposed to be based on science and evidence. Until research is conducted into the toxicity and persistence of the metabolites of pesticides and published in peer-reviewed journals, regulatory authorities have no peer-reviewed, science-based evidence to show that any of the current residue levels in food or in the environment are safe. Ignoring them or assuming that they are safe cannot be regarded as an acceptable regulatory practice. Regulatory authorities have a duty of care to ensure that the general population is not harmed by these toxic chemicals.

NOTES

[1] Food Standards Australia and New Zealand, "20th Australian Total Diet Survey."

[2] Short, *Quick Poison, Slow Poison*; Colborn, Dumanoski, and Myers, *Our Stolen Future*; Cadbury, *Feminization of Nature*; Cox, "Glyphosate (Roundup)"; Colborn, "A Case for Revisiting the Safety of Pesticides."

[3] D. W. Sparling, Gary Fellers. "Comparative Toxicity of Chlorpyrifos, Diazinon, Malathion and Their Oxon Derivatives to Larval *Rana boylii*," *Environmental Pollution* 147 (2007): 535–39.

[4] Colborn, "A Case for Revisiting the Safety of Pesticides."

[5] Cox, "Glyphosate (Roundup)."

[6] Elina Jerschow et al., "Dichlorophenol-Containing Pesticides and Allergies: Results from the U.S. National Health and Nutrition Examination Survey 2005–2006," *Annals of Allergy, Asthma & Immunology* 109, no. 6 (December 2012): 420–25.

[7] "U.S. President's Cancer Panel Annual Report," 2010.

[8] Bergman et al., *State of the Science of Endocrine Disrupting Chemicals 2012*.

[9] National Institute of Environmental Health Sciences, http://www.niehs.nih.gov (accessed July 15, 2013).

Reliable Regulatory Authorities

"Trust us — we have it all under control."

O ne of the greatest pesticide myths is that government regulatory authorities ensure that agricultural poisons are used safely. Their messages are that no adverse health effects occur when these chemicals are used as per "good agricultural practices" and when used in accordance with the directions on the label.

Some countries have or are about to have food safety regulations that require farmers to be trained in using chemicals and keeping records of their use. There are private or market-based food safety schemes with similar and in some cases stricter requirements.

However, most of the approximately 206 nations in the world (including observer and disputed states) do not have adequate regulatory systems that monitor whether agricultural poisons are used as directed by the label and as per "good agricultural practices." This is a major issue in the developing world where most farmers have a limited or no ability to read or write. In many cases they have to rely on the advice from local sales agents as to which pesticide to use, the rate it is mixed with water, and the amount to use on the crop.

Quite often the local sales agent has limited literacy and numeracy as well and does not have the technical knowledge to give this advice. In other cases they corruptly sell whatever pesticide they have in stock and give erroneous advice.

Most farmers in the developing world rarely use safety equipment like face masks or protective clothing and gloves. In some cases they mix the pesticide with water, stir it with their bare hands, and then splash it over the plants from a bucket with bare hands because they cannot afford spray equipment or protective clothing. In other cases this is done because the farmers do not understand that the chemicals are toxic to humans. They store the chemicals and mix them up in their small huts, surrounded by their family members and next to their food. Consequently the highest rates of pesticide poisonings are among farmers, their families, farm workers, and in rural communities in the developing world.

Most developed countries and some developing countries have monitoring programs to test the food being sold in shops and markets to detect if there are residues that exceed the maximum residue limits (MRLs). In most cases the residues are below the MRLs. The conclusion is that because most of the pesticide residues are below the MRLs the usage is safe, and consumers are safe because the food residues are within the acceptable safety margins.

However, as shown by the information in the previous chapters, these MRLs are highly questionable as they have been based on outdated methodologies that do not test for:
- Mixtures and cocktails of chemicals
- The actual formulated product
- The metabolites and impurities of pesticides
- The special requirements of the fetus and the newborn
- Endocrine disruption
- Intergenerational effects
- Developmental neurotoxicity

History shows that regulatory authorities have consistently failed to prevent the contamination of the environment and human health by products previously designated safe, such as asbestos, lead, mercury, dioxins, PCBs, DDT, dieldrin, and other persistent organic pollutants. In many cases these products are still widely used, such as mercury in tooth fillings, as a preservative in vaccines, and as a fungicide in agriculture. DDT is still widely used in countries like Uganda and China. Lead paint and white asbestos are still widely used in many countries. For the few products that have been withdrawn, it was decades after good scientific evidence was presented to demonstrate that they are harmful.

Regulatory authorities around the world are disregarding a large body of published science conducted by several hundred trained scientists and experts in these fields that clearly shows that the current methods of determining the safety of the agricultural poisons are grossly inadequate.

Dr. Theo Colborn gives examples that show that the EPA, the main regulatory authority in the United States, is ignoring a wealth of peer-reviewed scientific studies and is largely basing its conclusions on unpublished studies that have been commissioned by the pesticide industry. "Although this information is available, the U.S. EPA has rarely used the open literature in its risk assessments, generally using only data submitted by manufacturers."[1]

She states that by only relying on the data provided by pesticide manufacturers, the EPA is missing nearly all the delayed developmental, morphologic, and functional damage to the fetus. They are also missing data on the way pesticides interfere with the physiological systems in humans. "For example, Brucker-Davis (1998) published a comprehensive review of the open literature in which she found 63 pesticides that interfere with the thyroid system— a system known for more than a century to control brain development, intelligence, and behavior. Yet, to date, the U.S. EPA has never taken action on a pesticide because of its interference with the thyroid system."[2]

The WHO and UNEP meta-study clearly states that the current regulatory systems are inadequate when it comes to the issue of endocrine-disrupting chemicals. "We cannot be confident that the current system of protecting human and wildlife population from chemicals with endocrine activity is working as well as it should to help prevent adverse health impacts on human and wildlife populations."[3]

The USPCP report was critical about the current testing methodologies and the lack of action taken by regulatory authorities in reviewing the toxicity of chemicals based on the latest peer-reviewed science:

> The prevailing regulatory approach in the United States is reactionary rather than precautionary. That is, instead of taking preventive action when uncertainty exists about the potential harm a chemical or other environmental contaminant may cause, a hazard must be incontrovertibly demonstrated before action to ameliorate it is initiated. Moreover, instead of requiring industry or other proponents of specific chemicals, devices, or activities to prove their safety, the public bears the burden of proving that a given environmental exposure is harmful.[4]

REACH—THE EUROPEAN UNION'S COMPREHENSIVE REVIEW OF CHEMICALS

The European Union (EU) has started a major review of all widely used chemicals under a new regulation called the Registration, Evaluation, and Authorization of Chemicals (REACH).[5]

REACH started in 2007 and will be fully implemented by 2018. Initially the assessments were based on chemicals that are produced or imported in the EU of more than 1,000 metric tons per year. By 2018 it will cover chemicals that are in the order of one metric ton per year. About 143,000 chemical substances marketed in the European Union were pre-registered by the December 1, 2008, deadline.[6]

REACH will be missing several key tests, such as those for mixtures and cocktails of chemicals; the special requirements of fetuses, newborns, and growing children; endocrine disruption; pesticide metabolites; intergenerational effects; and developmental neurotoxicity. It will also allow products that are produced in the EU or imported in quantities of less than one metric ton to be exempt.

REACH is considered to be a good start in that this will be the first time in the world that there has been a comprehensive review of all the common chemicals that are used in a region. It will be the first time that a regulatory body will assess formulated products for their adverse effects instead of assessing only single active ingredients.

Instead of congratulating the EU for taking such a long overdue and important initiative and following the EU's example, a group of countries—including the United States, Brazil, Australia, India, Japan, Mexico, Singapore, South Africa, Thailand, Chile, Israel, South Korea, and Malaysia—put diplomatic pressure on the EU to have the regulation watered down on the basis that it would hamper the free trade in chemicals.

The prevailing regulatory approach in the United States is **REACTIONARY RATHER THAN PRECAUTIONARY.** Instead of requiring industry or other proponents of specific chemicals, devices, or activities to prove their safety, the public bears the burden of proving that a given environmental exposure is harmful.

EXAMPLES OF REGULATORY INACTION, NEGLECT, AND MISMANAGEMENT

Pesticides are used in our food production because they are toxic. Their primary role is to kill pests, diseases, and weeds by poisoning them. Regulatory authorities have a duty to ensure that humans and the environment are not being adversely affected by these poisons.

So far it has been the actions of civil society and concerned scientists that have made regulators take limited actions on the uses of some chemicals. Their intervention can sometimes result in modifications in how the chemicals are used or, in rare cases, a ban to improve the safety of humans and the environment. It should not be up to the public to spend enormous human and financial resources to prove that they are harmful. It should be up to the industry, the sector that profits from these chemicals, to prove unequivocally that they are safe when people and the environment are exposed to them.

Regulatory authorities should take a preventive approach as advocated by the USPCP. When a newly published peer-reviewed scientific study indicates a health issue with the current use patterns of a pesticide, the precautionary principle* should be invoked to ensure that there is no harm. This could mean that the use of the chemical be suspended until independently published, peer-reviewed scientific testing shows what level of exposure is safe.

Unfortunately the opposite approach is the reality, with regulatory authorities ignoring the new science and usually only taking action after years of work by the concerned sections of civil society and the scientific community.

GLYPHOSATE REGULATION—DECADES OF IGNORING SCIENTIFIC RESEARCH

The regulation of glyphosate is a good example of authorities ignoring an extensive body of published scientific study showing the harm that can be caused by this widely used pesticide. It is probably the most common herbicide used in the world, and its use is increasing due to the introduction of glyphosate-resistant GMO crops.

Glyphosate and its formulations are considered safe, so consequently they are widely used to spray roadsides, sidewalks, children's playgrounds, parks, and gardens as well as in food production. Commonly cited information arguing that glyphosate is very safe is found on the website for EXTOXNET, the Extension Toxicology Network. The site claims to be a source of "objective, science-based information about pesticides" and is a joint effort of the cooperative extension offices of Cornell University, Michigan State University, Oregon State University, and University of California-Davis. Major

---

* The precautionary principle is used in policymaking when there is a suspected risk to the health of humans and the environment and states that under these circumstances it may not be necessary to wait for scientific certainty to take protective action.

support and funding was provided by the USDA/Extension Service/National Agricultural Pesticide Impact Assessment Program. The primary files are maintained and archived at Oregon State University. However this information on glyphosate has not been revised since June 1996.[7] The cited studies were done in the 1980s. According to the EXTOXNET page on the Cornell University website, the information on glyphosate was reviewed by Monsanto, the manufacturer of glyphosate, in November 1992, which means that the scientific independence of the data has to be seriously questioned due to a large potential conflict of interest.[8]

EXTOXNET is gradually being replaced by fact sheets from the National Pesticide Information Center (NPIC), a cooperative agreement between Oregon State University and the U.S. Environmental Protection Agency. The general fact sheet on glyphosate was reviewed in September 2010 and states, "Glyphosate exposure has not been linked to developmental or reproductive effects in rats except at very high doses that were repeated during pregnancy. These doses made the mother rats sick. The rat fetuses gained weight more slowly, and some fetuses had skeletal defects."[9] The fact sheet gives the impression that glyphosate-based products are safe, though they may cause some problems "at very high doses."

In the case of Australia, as another example, the ADI for glyphosate of 0.3 milligrams per kilogram was based on a NOAEL that was established in February 14, 1985, by the regulatory Australian Pesticides and Veterinary Medicines Authority (APVMA). The WHO and the U.S. EPA have also set an ADI of 0.3 milligrams per kilogram. Numerous studies have been published in the nearly three decades years since this NOAEL was established that link glyphosates and glyphosate-based herbicides to a wide range of negative health effects.

CELL DAMAGE—PRECURSORS TO CANCER AND BIRTH DEFECTS

Research has shown that glyphosate can cause genetic damage, developmental disruption, morbidity, and mortality even at what are currently considered normal levels of use.[10] The article "Differential Effects of Glyphosate and Roundup on Human Placental Cells and

> **Glyphosate damaged human placental cells within EIGHTEEN HOURS of exposure, even at concentrations lower than those found in commercially available pesticides and herbicides.**

Aromatase," published by Richard et al. in *Environmental Health Perspectives*, revealed evidence that glyphosate damaged human placental cells within eighteen hours of exposure, even at concentrations lower than those found in commercially available pesticides and herbicides. The scientists stated that "this effect increases with concentration and time or in the presence of Roundup adjuvants."[11]

Researchers of a study published in the journal *Toxicology* studied four different commercial glyphosate formulations and observed breaks in 50 percent of the DNA strands in human liver cells at doses as low as five parts per million. This damage affects the way DNA sends messages to several physiological systems, including the endocrine system. The researchers stated that this is significant because the liver is the first detoxification organ and is sensitive to dietary pollutants.[12]

TERATOGENICITY (BIRTH DEFECTS) IN ANIMALS

Clements et al. published a study in 1997 showing damage to DNA in bullfrog tadpoles after exposure to glyphosate. The scientists concluded that glyphosate's "genotoxicity at relatively low concentrations" was of concern.[13] A 2003 study by Lajmanovich et al. found that up to 55 percent of tadpoles exposed to a glyphosate herbicide had deformities to the mouth, eyes, skull, vertebrae, and tails.[14] A 2003 study by Dallegrave et al. found that the offspring of pregnant rats dosed with glyphosate had increased skeletal abnormalities.[15]

A 2004 study conducted by biologists at Trent University, Carleton University (Canada), and the University of Victoria (Canada) showed that concentrations of several glyphosate herbicides at levels found in the environment caused developmental problems in tadpoles. The exposed tadpoles did not to grow to the normal size, took longer than normal to develop, and between 10 and 25 percent had abnormal sex organs.[16]

A 2010 study found that almost 60 percent of tadpoles treated with Roundup at one part per million had malformations such as kyphosis, scoliosis, and edema.[17] A 2012 study by Relyea found that tadpoles exposed to concentrations of Roundup found in the environment had changes to their tails.[18]

One of the most significant studies investigating the toxicity of both Roundup as a formulation and of glyphosate as the active ingredient, published by Alejandra Paganelli et al. in 2010, explains one of the ways they cause teratogenicity. The researchers found that both Roundup and glyphosate by itself caused severe malformations in the embryos of chickens and frogs and that this could occur in frogs when exposed to less than 0.5 parts per million. The researchers identified the specific mechanism that glyphosate and Roundup use to cause the malformations. They found that the chemicals disrupted a key biochemical mechanism, the retinoic acid signaling pathway.[19]

The retinoic acid signaling pathway is used by all vertebrates, including humans, to ensure the normal development of organs, bones and tissues in embryos. It is also essential for normal sexual development, especially in males. The pathway signals the exact time and place that the development of organs and tissues occurs in embryos. It also corrects malformations if they start. Disrupting its normal balance means that the various organs and tissues can be given signals to form incorrectly, and the pathway cannot correct any of these embryo malformations when they start forming.[20]

Glyphosate adversely affects the shikimate pathway in plants, interfering with synthesis of key compounds essential to their life. This effect is regarded as the main reason for its effectiveness as a broad-spectrum herbicide; however, there are studies that show other mechanisms in the way it affects plants, such as its ability to chelate key plant nutrients resulting in severe nutrient deficiencies. Because the shikimate pathway is only found in plants, it is assumed that glyphosate does not affect animals and therefore is safe; however, the retinoic acid signaling pathway in animals is very similar to the shikimate pathway.

Research by Mesnage et al. found that Roundup from 1 ppm to 20,000 ppm causes cells of the human body to die through necrosis. At 50 ppm it delays the cellular apoptosis that is essential for the normal functioning and regeneration of cells, body tissues, and organs.[21]

GLYPHOSATE AND CANCERS

A case-controlled study published in March 1999 by Swedish scientists Lennart Hardell and Mikael Eriksson showed that non-Hodgkin's lymphoma (NHL) is linked to exposure to a range of pesticides and herbicides, including glyphosate.[22] Prior to the 1940s, non-Hodgkin's lymphoma was one of the world's rarest cancers. Now it is one of the most common. Between 1973 and 1991, the incidence of non-Hodgkin's lymphoma increased at the rate of 3.3 percent per year in the United States, making it the third fastest-growing cancer.[23] In Sweden, the incidence of NHL has increased at the rate of 3.6 percent per year in men and 2.9 percent per year in women since 1958.

Several animal studies have shown that glyphosate can cause gene mutations and chromosomal aberrations. These types of genetic damage can be the precursors of cancer.[24]

A study published in 2004 found that glyphosate-based herbicides caused cell-cycle dysregulation, which leads to cancers. According to the researchers, "Cell-cycle dysregulation is a hallmark of tumor cells and human cancers. Failure in the cell-cycle checkpoints leads to genomic instability and subsequent development of cancers from the initial affected cell." The researchers tested several glyphosate-based pesticides and found that all of them caused cell-cycle dysregulation.[25]

Research (mentioned previously in chapter 2) published in the peer-reviewed scientific journal *Food and Chemical Toxicology* in 2013 found that glyphosate at residue levels commonly found in people caused a five- to thirteenfold increase in the multiplication of estrogen-sensitive human breast cancer cells.[26]

ENDOCRINE DISRUPTION

Thongprakaisang et al. found glyphosate acted like estrogen to cause the multiplication of estrogen-sensitive breast cancer cells.[27]

Gasnier et al. in 2009 reported endocrine-disrupting actions of glyphosate at 0.5 ppm. According to the authors this is "800 times lower than the level authorized in some food or feed (400 ppm, USEPA, 1998)."[28]

DISRUPTION OF METABOLIC PATHWAYS

One of the most significant studies was published by Samsel and Seneff in the peer-reviewed scientific journal *Entropy* in 2013. This comprehensive review, titled "Glyphosate's Suppression of Cytochrome P450 Enzymes and Amino Acid Biosynthesis by the Gut Microbiome: Pathways to Modern Diseases," showed how glyphosate disrupted numerous biochemical pathways within the human body, including gut microorganisms, and consequently could lead to numerous diseases.[29] The evidence in the scientific paper is voluminous and compelling and opens up significant areas where research is needed on glyphosate and other chemicals' potential to adversely affect human health.

Cytochrome P450 (CYP) is known as a superfamily of enzymes that are responsible for around 75 percent of metabolic reactions involved in drug metabolism and the oxidation of organic molecules and are present in most tissues of the body. They metabolize thousands of chemicals made by the body and those absorbed from external sources such foods, water, gut microorganisms, and the atmosphere. They are also involved in the metabolism and synthesis of numerous key biochemical compounds such as retinol, vitamin D, neurotransmitters such as serotonin and melatonin, and other compounds critical to health such as L-tryptophan and cholesterol. CYPs are very important in synthesis and metabolism of hormones such as estrogen, testosterone, aldosterone, androstenedione, cortisol, corticosterone, and dehydroepiandrosterone to ensure homeostasis. Very significantly CYPs metabolize toxic compounds such as pesticides, drugs, and other chemicals

Samsel and Seneff identified numerous ways that glyphosate disrupts the CYP enzymes and how this can cause many diseases. The authors state,

> Glyphosate's inhibition of cytochrome P450 (CYP) enzymes is an overlooked component of its toxicity to animals. CYP enzymes play crucial roles in biology, one of which is to detoxify xenobiotics. Thus, glyphosate enhances the damaging effects of other food borne chemical residues and environmental toxins. Negative impact on the body is insidious and manifests slowly over time as inflammation damages cellular systems throughout the body. Here, we show how interference with CYP enzymes acts synergistically with disruption of the biosynthesis of aromatic amino acids by gut bacteria, as well as impairment in serum sulfate transport. Consequences are most of the diseases and conditions associated with a Western diet, which include gastrointestinal disorders, obesity, diabetes, heart disease, depression, autism, infertility, cancer and Alzheimer's disease. We explain the documented effects of glyphosate and its ability to induce disease, and we show that glyphosate is the "textbook example" of exogenous semiotic entropy: the disruption of homeostasis by environmental toxins.[30]

Samsel and Seneff's research also confirms the earlier research by Paganelli et al. on glyphosate's ability to disrupt the retinoic acid pathway. As stated previously, retinoic acid has a key role in the development of embryos to stop birth defects (teratogenicity) from developing. One of the ways it does this is to metabolize excess retinol (a form of vitamin A). A key enzyme in retinoic acid pathway that does this is CYP26, one of the many members of the cytochrome P450 superfamily. The ability of glyphosate to disrupt this important group of enzymes is an example of one of the ways it disturbs the retinoic acid pathway.

The potential of numerous other chemicals, including pesticides, to adversely affect the cytochrome P450 pathways in all living biota, including humans, needs to be actively researched given that these pathways are responsible for a significant percentage of metabolic functions. At this stage research is in its infancy, and the complexity of these interactions is not adequately understood due to being mostly overlooked. The body of evidence presented by Samsel and Seneff shows that disruption of these key metabolic systems by xenobiotics such as synthetic chemicals is one of the reasons for the dramatic increases of a large range of diseases, especially in developed countries and the growing middle classes in developing countries.

> Glyphosate caused severe **MALFORMATIONS IN THE EMBRYOS OF CHICKENS AND FROGS,** and this could occur in frogs when exposed to less than 0.5 parts per million.

This finding is reinforced by the significant number of studies documenting a range of birth defects, cancers, and other diseases linked to the exposure of glyphosate.

DISRUPTION OF THE GUT MICROBIOME

Samsel and Seneff's paper identified how glyphosate disrupted the gut microbiome, causing the suppression of biosynthesis of cytochrome P450 enzymes and key amino acids. In a later paper, "Glyphosate, Pathways to Modern Diseases II: Celiac Sprue and Gluten Intolerance," Samsel and Seneff showed how the current increase in celiac disease and gluten intolerance in people was linked to glyphosate's adverse effects on the gut microbiome. They highlighted that glyphosate is patented as a biocide, and consequently it kills the beneficial gut bacteria, leading to a rise in intestinal diseases.[31] Krüger et al. showed that glyphosate has this effect in the microbiome of horses and cows.[32] Shehata et al. found the same effects in poultry; the researchers state, "Highly pathogenic bacteria as *Salmonella* Entritidis, *Salmonella* Gallinarum, *Salmonella* Typhimurium, *Clostridium perfringens* and *Clostridium botulinum* are highly resistant to glyphosate. However, most of beneficial bacteria as *Enterococcus faecalis*, *Enterococcus faecium*, *Bacillus badius*, *Bifidobacterium adolescentis* and *Lactobacillus* spp. were found to be moderate to highly susceptible."[33] Both groups of researchers postulated that glyphosate is associated with the increase in botulism-mediated diseases in these domestic farm animals.

KIDNEY DISEASE

Since the 1990s, researchers in Sri Lanka have reported massive kidney failure in rice paddy workers exposed to glyphosate in combination with minerals in hard water. According to Jayasumana et al., glyphosate's strong chelating properties allow it to combine with heavy metals and arsenic in hard waters, resulting in damage to renal tissues and thereby causing chronic kidney diseases. The authors concluded that, "The GMA [glyphosate-metal/arsenic complex] lattice hypothesis gives rational and consistent explanations to the many observations and unanswered questions associated with the mysterious kidney disease in rural Sri Lanka. Furthermore, it may explain the similar epidemics of CKDu [chronic kidney disease of unknown etiology] observed in Andra Pradesh, India and Central America."[34]

OXIDATIVE STRESS AND CELL DAMAGE

Oxidative stress is an imbalance between free radicals and the body's ability to repair the damage caused by free radicals. It has been linked to Alzheimer's, cancer, and Parkinson's disease, among other health issues. Cattani et al. found that both acute and chronic exposure to Roundup induced oxidative stress resulting in neural cell death and neurotoxic effects in the hippocampus region of the brain in immature rats.[35] Lushchak et al. found that a ninety-six-hour exposure to low levels of Roundup in water caused oxidative stress to the cells in the brains, livers, and kidneys of goldfish.[36]

Studies by El-Shenawy and de Liz Oliveira Cavalli et al. confirm that Roundup and glyphosate caused oxidative stress and necrosis in cells, including the liver, testis, and Sertoli cells in rats.[37]

REGULATORY RESPONSES NOT CONSISTENT WITH THE PUBLISHED SCIENCE

The Australian regulators have not revised the NOAEL that they set in 1985, even though there is an enormous body of scientific research linking glyphosate and glyphosate-formulated products to a wide range of negative health effects. This failure to act is shared by many other regulatory authorities around the world.

A greater concern was the U.S. EPA's decision in May 2013 to significantly increase the allowed MRLs of glyphosate in food and other crops in response to a petition prepared by Monsanto, despite the large number of peer-reviewed studies showing multiple health problems associated with exposure to glyphosate. Unfortunately, this should not come as a surprise, as Dr. Colborn has stated, "The U.S. EPA has rarely used the open literature in its risk assessments, generally using only data submitted by manufacturers."[38]

Monsanto petitioned for the MRL increase because the large number of glyphosate-tolerant GMO crops has resulted in a major increase in the amount of glyphosate used in farming. As a result, glyphosate residues began to exceed the existing MRLs. Rather than risking farmers finding a way to reduce the glyphosate usage in agriculture in accordance with the law and possibly losing sales, Monsanto sought to change the law by increasing these MRLs so that testing will show that glyphosate residues are below the MRLs and therefore being used "safely."

Research conducted by Dr. Nancy Swanson and colleagues shows a link between the increase in the use of glyphosate, the acres of land under GMOs, and a range of diseases in the United States.[39]

Dr. Swanson used statistical data from credible sources such as the Centers for Disease Control and Prevention and the United States Department of Agriculture and turned them into graphs that showed the correlations. To ensure the accuracy of the correlations they were adjusted using a standard statistical method of analyzing data called Pearson's correlation coefficients.

The Pearson's correlation coefficient is used by statisticians to determine how closely two data sets are correlated. To understand the following graphs, the letter R represents the value of the correlation. If R = +.70 or higher it means that there is a very strong positive relationship, +.40 to +.69 indicates a strong positive relationship, +.30 to +.39 a moderate positive relationship, +.20 to +.29 a weak positive relationship, and +.01 to +.19 means no or a negligible relationship.

These numbers can be used by statisticians calculate the probability that a relationship between two variables could have happened by chance. A statistically significant finding is one that is determined to be very unlikely to happen by chance. In the following graphs, the letter p is the abbreviation for the probability. For example, if p = .05 or 5 percent, it means that there is less than a one in twenty chance that the observed relationship could have happened by chance, and the findings are designated as significant. Likewise, p = .01 or 1 percent means that there is less than a one in one hundred chance and thus the findings are designated as highly significant.[40]

The diabetes graph is particularly relevant; the trend line shows that cases were slowly decreasing until the increase in the use of glyphosate. At this point the new cases of diabetes rapidly increase at the same rate as the increase of the use of glyphosate.

The Pearson's correlation coefficients show a strong relationship between the rise of these diseases and glyphosate application, and these findings are consistent with the numerous scientific studies linking it to a range of diseases. This presents a strong body of evidence that glyphosate regulation is seriously inadequate in protecting human health.

The graphs show sudden increases in the rates of diseases that start in the mid-1990s. These changes coincide with the commercial production of GMO crops, particularly the glyphosate-resistant crops. Since the mid-1990s the use of no other pesticide has

NUMBER OF CHILDREN (6-21 YRS)
WITH AUTISM SERVED BY IDEA

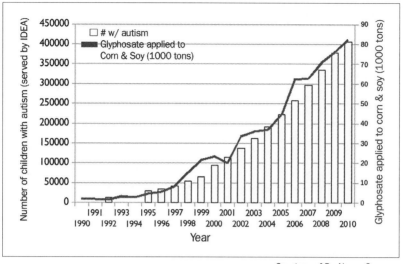

Courtesy of Dr. Nancy Swanson

DEATHS DUE TO THYROID CANCER (ICD C93 &193)

Plotted against %GE corn & soy (R=0.876, p<=7.947e-05) and glyphosate applied to corn & soy (R=0.9583, p<=2.082e-08)

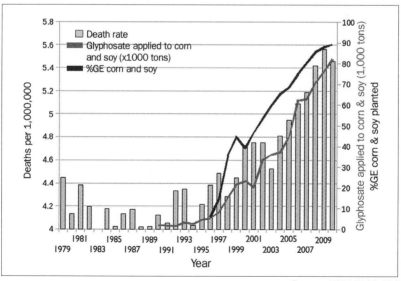

Sources: USDA:NASS; CDC

AGE ADJUSTED DEATHS DUE TO OBESITY (ICD E66 & 278)

Plotted against %GE corn & soy (R=0.9618, p<=3.504e-06) and glyphosate applied to corn & soy (R=0.9616, p<=1.695e-08)

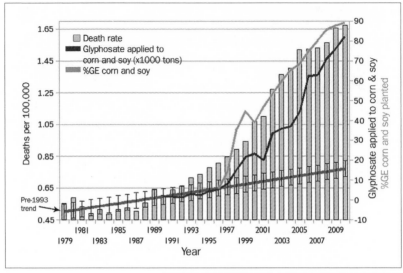

Sources: USDA:NASS; CDC, Courtesy of Dr. Nancy Swanson

DEATHS FROM ALZHEIMER'S (ICD G30.9 & 331.0)

Plotted against glyphosate use (R=0.9319, p<=9.903e-08) and %GE corn & soy (R=0.9511, p<=5.51e-06)

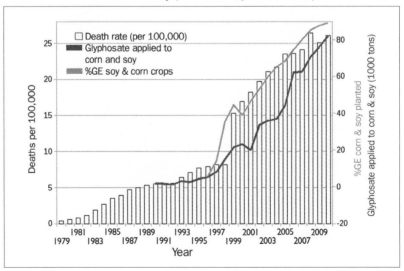

Sources: USDA:NASS; CDC, Courtesy of Dr. Nancy Swanson

ANNUAL INCIDENCE OF DIABETES (AGE ADJUSTED)

Plotted against %GE corn & soy crops planted (R=0.9547, p<=1.978e-06) along with glyphosate applied to corn & soy in U.S. (R=0.935, p<=8.303e-08).

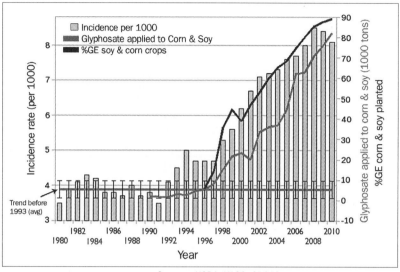

Sources: USDA: NASS; CDC, Courtesy of Dr. Nancy Swanson

URINARY/BLADDER CANCER INCIDENCE (AGE ADJUSTED)

Plotted against % GE corn and soy (R=0.9449, p<=7.1e-06) and glyphosate applied to corn and soy (R=0.981, p<=4.702e-09)

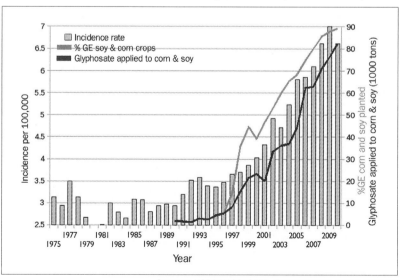

Sources: USDA:NASS; SEER, Courtesy of Dr. Nancy Swanson

DEATHS DUE TO MULTIPLE SCLEROSIS (ICD G35 & 340)
Plotted against percentage of GE soy & corn (R=0.9477, p<=6.339e-06)
and glyphosate applied to soy & corn (R=0.9005, p<=5.079e-07)

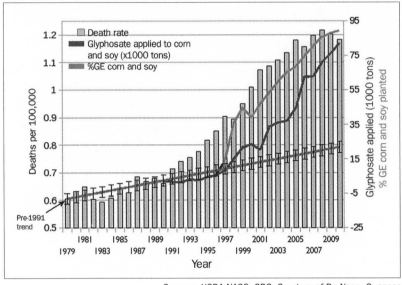

Sources: USDA:NASS; CDC, Courtesy of Dr. Nancy Swanson

DEATHS DUE TO INTESTINAL INFECTION (ICD A04, A09; 004, 009)
Plotted against glyphosate applied to corn and soy
(R=0.9762, p<=6.494e-09)

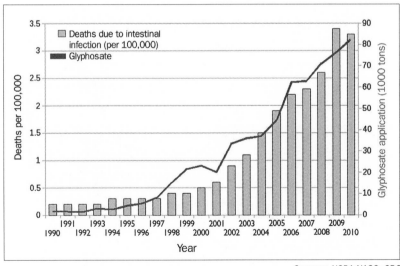

Sources: USDA:NASS; CDC

LIVER AND INTRAHEPATIC BILE DUCT CANCER INCIDENCE (AGE ADJUSTED)

Plotted against glyphosate applied to corn & soy
(R=0.9596, p<=4.624e-08) along with %GE corn & soy planted
in U.S. (R=0.9107, p<=5.402e-05)

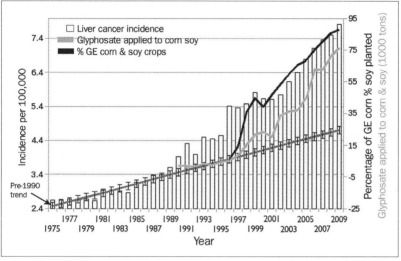

increased as much as glyphosate, and this large uptick in use in the United States is mostly due to the expansion of glyphosate-resistant genetically engineered crops.

EUROPEAN REGULATORS AND GLYPHOSATE

In general the countries of the European Union tend to review pesticides more frequently than other countries in the world, but the way some of these reviews are conducted needs to be seriously questioned. The EU regulators' reaction to the Paganelli study, which found that glyphosate caused severe malformations in the embryos of chickens and frogs, provides an insight into how some of the review processes work.

The regulators and the pesticide industry strongly rebutted the Paganelli et al. study rather than trying to repeat it to see if they could get the same results, even though repeating a scientific study by following its material and methods is regarded as the correct way to check the results for accuracy and consistency. The European regulators instead used other studies to invalidate the Paganelli study, claiming there was no evidence that glyphosate causes birth defects, a move that raises concerns about how regulators and the pesticide industry work together to ensure that toxic products are still sold and widely used despite a strong body of science showing these products should be severely restricted or banned. The regulators' assessment minimized the evidence presented by Paganelli et al., claiming that the exposure conditions used by the researchers were "highly artificial" and that there was "no clear-cut link" between birth defects and heavy pesticide use.[41]

In response, Michael Antoniou and a team of researchers published a study in the journal of *Environmental and Analytical Toxicology* in 2012 subtitled "Divergence of Regulatory Decisions from Scientific Evidence," documenting how the German Federal Office of Consumer Protection and Food Safety (BVL) not only ignored an extensive body of published science, they actively misrepresented the data in the selective studies they used to set the acceptable daily intake.[42] The BVL is responsible for making the recommendations on the safety of glyphosate for all the countries in the European Union. The independent scientists who reviewed the studies used by BVL found that there was more evidence that glyphosate caused malformations, not less as stated by the German regulators. The independent scientists concluded:

> However, examination of the German authorities' draft assessment report on the industry studies, which underlies glyphosate's EU authorisation, revealed further evidence of glyphosate's teratogenicity [ability to cause birth defects]. Many of the malformations found were of the type defined in the scientific literature as associated with retinoic acid teratogenesis. Nevertheless, the German and EU authorities minimized these findings in their assessment and set a potentially unsafe acceptable daily intake (ADI) level for glyphosate.[43]

Antoniou et al. concluded that there was a need for a new review of the data by independent scientists to ensure a credible regulatory decision that will guarantee that people do not suffer adverse health effects from the permitted uses of glyphosate.[44]

UNPUBLISHED INDUSTRY-SPONSORED STUDIES

The German authorities' rebuttal of the Paganelli et al. study raises the issue of the use of unpublished non-peer-reviewed, industry-sponsored studies by regulatory authorities. The BVL's negations relied partly on unpublished industry studies commissioned for regulatory purposes. These studies made the claim that glyphosate did not cause birth defects or act as a reproductive toxin.

Regulatory authorities tend to use these studies to make their assessments instead of peer-reviewed studies that are published in scientific journals.

A high proportion of these unpublished industry studies are not available for review and assessments by other scientists and stakeholders because they are regarded as commercial-in-confidence. In many cases the reports and the toxicological data can only be obtained through court cases and/or the relevant discovery and freedom of information legislations in their respective countries.

This was the case with the review of the German authority's assessment of glyphosate by Antoniou et al. While the scientists managed to obtain the assessment report on glyphosate that the German authorities compiled in 1998, the industry toxicological data that was summarized in the assessment report was not publicly available. This data was claimed to be commercially confidential by the chemical company. In order to get this data the Pesticide Action Network Europe took legal action through the courts.[45]

THE NEED FOR TRANSPARENT, PUBLISHED, PEER-REVIEWED SCIENCE

Regulatory authorities using unpublished, non-peer-reviewed, industry-sponsored studies should be seen as a major problem in the current regulatory decision-making processes. As well as the issue of potential conflicts of interest, there is also the fact that the public has the right to know about the research used to determine the safety of the pesticides regularly found in their food.

Good science is based on peer-reviewed papers that are published in credible journals. These papers should clearly document the materials and methods used so that other scientists can accurately repeat the research to see if they get can consistently get the same results. This will confirm the veracity and the credibility of the research.

These papers should be reviewed by independent experts in the field for an unbiased and critical analysis of the research. One of the purposes of peer review is to ensure that the researchers have not overlooked any aspects in the design of experiments and the interpretation of data that may have influenced the outcomes of the research and the resulting conclusions. Studies published in peer-reviewed journals are available for all the relevant stakeholders to read and analyze, which ensures that the process is transparent and allows a wider and more critical debate over the data and conclusions.

Involved researchers should also declare any potential conflicts of interests, such as if they are paid by the industry to do the study. This transparency helps to overcome the issue of hidden agendas and bias and should be seen as an essential part of a fully open process. There is a lot of literature documenting or alleging that the industry has manipulated data to ensure favorable regulatory outcomes. Removing the secrecy and ensuring that all the decisions are made on the basis of the transparent processes that come with published peer-reviewed studies will reduce, if not eliminate, allegations of industry bias and interference. Unfortunately this is not the case with the current practices.

INCONSISTENCIES OF VARIOUS REGULATORY AUTHORITIES

Another important reason for ensuring that regulatory decisions are made on the basis of published, peer-review science is to remove the many inconsistencies in the decisions made by regulatory authorities. It's senseless that the residue levels of chemical formulas that one country deems unsafe for use, another country deems safe and legal, even though what's dangerous for humans in one country should be just as dangerous elsewhere.

Chlorpyrifos has been banned from use in the production of food in the United States, but most countries have ignored the scientific studies and continue to allow its widespread use. DDT is still widely used in countries like India, China, and Uganda, despite the overwhelming body of scientific evidence indicating the numerous health and environmental problems that it causes and despite the fact that there are numerous safer alternatives.

Atrazine is one of the most commonly used herbicides globally; however, it was banned in the European Union and Switzerland because it had polluted most water sources, was detected in a substantial percentage of rain samples, and caused a wide range of negative health effects.

The Australian regulatory processes are good examples of these inconsistencies. A review published by the World Wildlife Fund and the National Toxics Network showed that Australian farmers use around eighty chemicals that were banned in other countries because they pose risks to human health and the environment. The authors state, "The list includes 17 chemicals that are known, likely or probable carcinogens, and 48 chemicals flagged as having the potential to interfere with hormones. More than 20 have been classified as either extremely or highly hazardous by the World Health Organisation yet remain available for use on Australian farms."[46]

IT'S SENSELESS that the residue levels of chemical formulas that one country deems unsafe for use, another country deems safe and legal, even though what's dangerous for humans in one country should be just as dangerous elsewhere.

Australia is usually among the last countries in the world to ban toxic chemicals, sometimes decades after publication of peer-reviewed science that other countries used as the basis to withdraw these chemicals from food production. Endosulfan is a good example. Australia was one of the last countries in the world to withdraw it from use. Also, when the U.S. EPA ended all food uses of chlorpyrifos after studies linked it to a range of negative health outcomes, the Australian regulator decided that there was no need to act on these studies and make similar changes to the way chlorpyrifos is used.

These inconsistencies show that the current processes of making regulatory decisions are more about political debates rather than logical decisions made on credible published science. They also confirm that the basis of most regulatory decisions is a reactionary rather than precautionary approach to chemical regulation.

THE NEED FOR A NEW REGULATORY APPROACH

There is an overwhelming body of scientific and other evidence showing that the current pesticide regulatory systems are not sufficient to ensure that pesticide use is safe for humans and the environment.

The USPCP clearly states that regulatory agencies are failing in their responsibilities to prevent the public from contracting serious illnesses such as cancer from exposures to environmental toxins such as pesticides. "In large measure, adequate environmental health regulatory agencies and infrastructures already exist, but agencies responsible for promulgating and enforcing regulations related to environmental exposures are failing to carry out their responsibilities. . . . A precautionary, prevention-oriented approach should replace current reactionary approaches to environmental contaminants in which human harm must be proven before action is taken to reduce or eliminate exposure."[47]

Regulators are ignoring a huge body of credible, peer-reviewed scientific studies and are instead making most of their decisions based on unpublished industry studies. This approach is clearly flawed and strongly biased toward industry over independent scientists and researchers.

Dr. Theo Colborn wrote in *A Case for Revisiting the Safety of Pesticides*: "In conclusion, an entirely new approach to determine the safety of pesticides is needed. It is evident that contemporary acute and chronic toxicity studies are not protective of future generations. . . . To protect human health, however, a new regulatory approach is also needed that takes into consideration this vast new knowledge about the neurodevelopmental effects of pesticides, not allowing the uncertainty that accompanies scientific research to serve as an impediment to protective actions."[48]

NOTES

[1] Colborn, "A Case for Revisiting the Safety of Pesticides."

[2] Ibid.

[3] Bergman et al., *State of the Science of Endocrine Disrupting Chemicals 2012.*

[4] "U.S. President's Cancer Panel Annual Report," 2010.

[5] European Commission, "What is REACH?"

[6] Ibid.

[7] EXTOXNET: The Extension Toxicology Network, University of California-Davis, Oregon State University, Michigan State University, Cornell University, and the University of Idaho, 1996, http://extoxnet.orst.edu/pips/glyphosa.htm (accessed January 26, 2014).

[8] "Glyphosate," EXTOXNET, University of California-Davis, Oregon State University, Michigan State University, Cornell University, and the University of Idaho, May 1994, http://pmep.cce.cornell.edu/profiles/extoxnet/dienochlor-glyphosate/glyphosate-ext.html (accessed January 24, 2014).

[9] "Glyphosate General Fact Sheet," National Pesticide Information Center, September 2010, http://npic.orst.edu/factsheets/glyphogen.pdf (accessed January 26, 2014).

[10] Cox, "Glyphosate (Roundup)."

[11] Richard et al., "Differential Effects of Glyphosate and Roundup."

[12] Céline Gasnier et al., "Glyphosate-Based Herbicides are Toxic and Endocrine Disruptors in Human Cell Lines," *Toxicology* 262 (2009): 184–91.

[13] Chris Clements, Steven Ralph, and Michael Petras, "Genotoxicity of Select Herbicides in *Rana catesbeiana* Tadpoles Using the Alkaline Single-Cell Gel DNA Electrophoresis (Comet) Assay," *Environmental and Molecular Mutagenesis* 29, no. 3 (1997): 277–88.

[14] Rafael C. Lajmanovich, M. T. Sandoval, Paola M. Peltzer, "Induction of

Mortality and Malformation in *Scinax nasicus* Tadpoles Exposed to Glyphosate Formulations," *Bulletin of Environmental Contamination Toxicology* 70, no. 3 (March 2003): 612–18.

[15] Eliane Dallegrave et al., "The Teratogenic Potential of the Herbicide Glyphosate-Roundup in Wistar Rats," *Toxicology Letters* 142, nos. 1–2 (April 2003): 45–52.

[16] Christina M. Howe et al., "Toxicity of Glyphosate-Based Pesticides to Four North American Frog Species," *Environmental Toxicology and Chemistry* 23, no. 8 (August 2004): 1928–38.

[17] Uthpala A. Jayawardena et al., "Toxicity of Agrochemicals to Common Hourglass Tree Frog (*Polypedates cruciger*) in Acute and Chronic Exposure," *International Journal of Agriculture and Biology* 12 (2010): 641–48.

[18] Rick A. Relyea, "New Effects of Roundup on Amphibians: Predators Reduce Herbicide Mortality; Herbicides Induce Antipredator Morphology," *Ecological Applications* 22 (2012): 634–47.

[19] Alejandra Paganelli et al., "Glyphosate-Based Herbicides Produce Teratogenic Effects on Vertebrates by Impairing Retinoic Acid Signaling," *Chemical Research in Toxicology* 23, no. 10 (August 2010): 1586–95.

[20] Ibid.

[21] Mesnage, Bernay, and Séralini, "Ethoxylated Adjuvants of Glyphosate-Based Herbicides."

[22] Lennart Hardell and Mikael Eriksson, "A Case-Control Study of Non-Hodgkin Lymphoma and Exposure to Pesticides," *Cancer* 85, no. 6 (March 15, 1999): 1353–60.

[23] Angela Harras, ed., *Cancer Rates and Risks*, 4th ed. (Washington, D.C.: U.S. Department of Health and Human Services, Public Health Service, National Institutes of Health, 1996).

[24] Cox, "Glyphosate (Roundup)."

[25] Julie Marc, Odile Mulner-Lorillon, and Robert Bellé, "Glyphosate-Based Pesticides Affect Cell Cycle Regulation," *Biology of the Cell* 96, no. 3 (April 2004): 245–49.

[26] Thongprakaisang et al., "Glyphosate Induces Human Breast Cancer Cells Growth."

[27] Ibid.

[28] Ibid.

[29] Anthony Samsel and Stephanie Seneff, "Glyphosate's Suppression of Cytochrome P450 Enzymes and Amino Acid Biosynthesis by the Gut Microbiome: Pathways to Modern Diseases," *Entropy* 15, no. 4 (2013): 1416–63.

[30] Ibid.

[31] Anthony Samsel and Stephanie Seneff, "Glyphosate, Pathways to Modern Diseases II: Celiac Sprue and Gluten Intolerance," *Interdisciplinary Toxicology* 6, no. 4 (2013): 159–84, http://sustainablepulse.com/wp-content/uploads/2014/02/Glyphosate_II_Samsel-Seneff.pdf (accessed March 21, 2014).

[32] Monika Krüger, Awad Ali Shehata, Wieland Schrödl, and Arne Rodloff, "Glyphosate Suppresses the Antagonistic Effect of *Enterococcus* spp. on *Clostridium botulinum*," *Anaerobe* 20 (April 2013): 74–78.

[33] Awad Ali Shehata, Wieland Schrödl, Alaa A. Aldin, Hafez M. Hafez, and Monika Krüger, "The Effect of Glyphosate on Potential Pathogens and Beneficial Members of Poultry Microbiota in Vitro," *Current Microbiology* 66, no. 4 (2012): 350–58.

[34] Channa Jayasumana, Sarath Gunatilake, and Priyantha Senanayake, "Glyphosate, Hard Water and Nephrotoxic Metals: Are They the Culprits Behind the Epidemic of Chronic Kidney Disease of Unknown Etiology in Sri Lanka?," *International Journal of Environmental Research and Public Health* 11, no. 2 (February 2014): 2125–47.

[35] Daiane Cattani et al., "Mechanisms Underlying the Neurotoxicity Induced by Glyphosate-Based Herbicide in Immature Rat Hippocampus: Involvement of Glutamate Excitotoxicity," *Toxicology* 320 (March 2014): 34–45.

[36] Oleh V. Lushchak et al., "Low Toxic Herbicide Roundup Induces Mild Oxidative Stress in Goldfish Tissues," *Chemosphere* 76, no. 7 (2009): 932–37.

[37] Nahla S. El-Shenawy, "Oxidative Stress Responses of Rats Exposed to Roundup and Its Active Ingredient Glyphosate," *Environmental Toxicology and Pharmacology* 28, no. 3 (November 2009): 379–85; Vera Lúcia de Liz Oliveira Cavalli et al., "Roundup Disrupted Male Reproductive Functions By Triggering Calcium-Mediated Cell Death In Rat Testis And Sertoli Cells," *Free Radical Biology & Medicine* 65 (December 2013): 335–46.

[38] Colborn, "A Case for Revisiting the Safety of Pesticides."

[39] Nancy Swanson, "Genetically Modified Organisms and the Deterioration of Health in the United States," *Sustainable Pulse*, April 27, 2013, http://sustainablepulse.com/2013/04/27/dr-swanson-gmos-and-roundup-increase-chronic-diseases-infertility-and-birth-defects (accessed August 24, 2013).

[40] Ibid.

[41] German Federal Office of Consumer Protection and Food Safety regulators (BVL), http://www.powerbase.info/images/b/b8/BVL2010.comments.Paganelli.pdf.

[42] Michael Antoniou et al., "Teratogenic Effects of Glyphosate-Based Herbicides: Divergence of Regulatory Decisions from Scientific Evidence," *Journal of Environmental and Analytical Toxicology* (2012): S4:006. doi:10.4172/2161-0525.S4-006.

[43] Ibid.

[44] Ibid.

[45] Ibid.

[46] Jo Immig, "A List of Australia's Most Dangerous Pesticides," National Toxics Network, July 2010, http://www.ntn.org.au/wp/wpcontent/uploads/2010/07/FINAL-A-list-of-Australias-most-dangerous-pesticides-v2.pdf.

[47] "U.S. President's Cancer Panel Annual Report," 2010.

[48] Colborn, "A Case for Revisiting the Safety of Pesticides."

Pesticides are Essential to Farming

"We will starve to death without pesticides."

T he greatest of all the myths is that we must be exposed to numerous toxic chemicals; otherwise we will have mass starvation. This myth states that it is impossible to grow enough food without the widespread use of these poisons.

The industry, both manufacturers and conventional farming organizations, and regulators consistently argue that not using these pesticides would cause crop failures and dramatic reductions in yields.

The main Australian pesticide regulator, the Australian Pesticides and Veterinary Medicines Authority (APVMA), is a good example of a regulator justifying the use of pesticides: "Pesticides and veterinary medicines are vital to quality food and fibre production. Australia's primary production is worth an estimated $30 billion a year with an export value of over $25 billion. Many primary producers rely on pesticides and veterinary medicines to protect their crops and animals from disease and pests."[1]

When pesticides are being reviewed by regulators for adverse effects to human health and the environment, the industry groups always warn that they have no alternative but to use these toxic chemicals as crop protection tools as the justification for not banning them. In the final outcome, it is usually business as usual, or regulators may decide to modify the way pesticides are used to lessen some negative impacts. Rarely are they withdrawn from use to ensure no adverse impacts on human health and the environment.

Trillions of dollars have been spent on research into conventional agriculture while at the same time in the last hundred years there has been an almost total neglect of research into organic agriculture. A significant proportion of this research funding has been to develop and test the efficacy of synthetic toxic chemicals as pesticides such as herbicides, insecticides, and fungicides.

Some comparison meta-studies, such as the recent ones published in *Nature* and *Agricultural Systems*, suggest that, on average, organic yields are 80 percent of conventional yields.[2] On the other hand, a meta-study by Badgley et al. suggests that the average organic yields are slightly below the chemical intensive yields in the developed world and higher than the conventional average in the developing world.[3] Assuming that the analyses in the journals *Nature* and *Agricultural Systems* are correct, 80 percent is an incredibly small yield gap in relation to the enormous level of research and resources that have been spent to achieve it.

The surprising fact is that millions of organic farmers have worked out how to get reasonable yields without the assistance of scientific research or the regular extension services that conventional agriculture receives.

The main reason for the lower yields in some organic systems has been the fact that research and development into organic systems has been largely ignored. U.S. $52 billion is spent annually on agriculture research worldwide. Less than 0.4 percent (four dollars in every thousand) is spent on solutions specific for organic farming systems.[4]

Yet despite this lack of funding, all the data sets from the global meta comparison studies have examples of organic systems that have the same or higher yields than conventional agriculture.

EXAMPLES OF HIGH-YIELDING ORGANIC SYSTEMS

The following examples of high-yielding organic systems show that under the right conditions organic farming systems can have equal or higher yields than chemical intensive farming systems.

U.S. Agricultural Research Service (ARS) Pecan Trial—The ARS organically managed pecans out-yielded the conventionally managed, chemically fertilized orchard in each of the past five years. Yields at the ARS organic test site surpassed the conventional orchard by eighteen pounds of pecan nuts per tree in 2005 and by twelve pounds per tree in 2007.[5]

The Wisconsin Integrated Cropping Systems Trials—The Wisconsin Integrated Cropping Systems Trials found that organic yields were higher in drought years and the same as conventional in normal weather years. In years with wet weather in the spring the organic yields can suffer when mechanical cultivation of weeds is delayed, and yields were found to be 10 percent lower. This could be corrected by using steam or vinegar for weed control, rather than tillage. The researchers attributed the higher yields in dry years to the organic soil's ability to take in rainfall more quickly. This is due to the higher levels of organic carbon, making the soils more friable and better able to store and capture rain.[6]

> The Wisconsin Integrated Cropping Systems Trials found that **ORGANIC YIELDS WERE HIGHER** in drought years and the same as conventional in normal weather years.

Scientific Review by Cornell University into Twenty-Two-Year-Long Rodale Field Study—The scientific review found:
- The improved soil allowed the organic land to generate yields equal to or greater than the conventional crops after five years.

- The conventional crops collapsed during drought years.
- The organic crops fluctuated only slightly during drought years, due to greater water-holding capacity in the enriched soil.
- The organic crops used 30 percent less fossil energy inputs than the conventional crops.[7]

Rodale Organic Low-/No-Till—The Rodale Institute has been trialing a range of organic low-tillage and no-tillage systems. The 2006 trials resulted in organic yields of 160 bushels an acre (bu/ac) compared to the Berks County average nonorganic corn yield of 130 bu/ac and the regional average of 147 bu/ac.

> The average corn yield of the two organic no-till production fields was 160 bu/ac, while the no-till research field plots averaged 146 bu/ac over 24 plots. The standard-till organic production field yielded 143 bu/ac, while the Farming Systems Trial's (FST's) standard-till organic plots yielded 139 bu/ac in the manure system (which received compost but no vetch N inputs) and 132 bu/ac in the legume system (which received vetch but no compost). At the same time, the FST's non-organic standard-till field yielded 113 bu/ac.[8]

Iowa Trials—The results from the Long Term Agroecological Research (LTAR), a twelve-year collaborative effort between producers and researchers led by Dr. Kathleen Delate of Iowa State University, shows that organic systems can have equal to higher yields than conventional systems. Consistent with several other studies, the data showed that while the organic systems had lower yields in the beginning, by year four they started to exceed the conventional crops. Across all rotations, organic corn harvests averaged 130 bushels per acre while conventional corn yield was 112 bushels per acre. Similarly, organic soybean yield was 45 bushels per acre compared to the conventional yield of 40 bushels per acre in the fourth year. Cost-wise, on average, the organic crops' revenue was twice that of conventional crops due to the savings afforded by not using chemical fertilizers and pesticides and the produce receiving better prices.[9]

MASIPAG Philippines—A research project conducted in the Philippines by MASI-PAG found that the yields of organic rice were similar to conventional systems.[10]

Other Examples—Professor George Monbiot, in an article in the *Guardian* in 2000, wrote that for the past 150 years wheat grown with manure has produced consistently higher yields than wheat grown with chemical nutrients in trials in the United Kingdom.[11]

> Dr. Welsh's study showed that
>
> ## ORGANIC AGRICULTURE PRODUCED BETTER YIELDS
>
> than conventional agriculture in adverse weather events, such as droughts or higher-than-average rainfall.

The study into apple production conducted by Washington State University compared the economic and environmental sustainability of conventional, organic, and integrated growing systems in apple production and found similar yields. "Here we report the sustainability of organic, conventional and integrated apple production systems in Washington State from 1994 to 1999. All three systems gave similar apple yields."[12]

In an article published in the peer-reviewed scientific journal *Nature*, Laurie Drinkwater and colleagues from the Rodale Institute showed that organic farming had better environmental outcomes as well as similar yields of both products and profits when compared to conventional, intensive agriculture.[13]

Dr. Rick Welsh of the Henry A. Wallace Institute reviewed numerous academic publications comparing organic production with conventional production systems in the United States. The data showed that the organic systems were more profitable. This profit was due not always to premiums but to lower production and input costs as well as more consistent yields. Dr. Welsh's study also showed that organic agriculture produced better yields than conventional agriculture in adverse weather events, such as droughts or higher-than-average rainfall.[14]

Nicolas Parrott of Cardiff University in the UK authored a report, titled *The Real Green Revolution*, in which he relates case studies that confirm the success of organic and agroecological farming techniques in the developing world. Average cotton yields on farms participating in the organic Maikaal Bio-Cotton Project are 20 percent higher than on neighboring conventional farms in the State of Madhya Pradesh in India. The System of Rice Intensification has increased yields from the usual two to three tons per hectare to yields of six, eight, or ten tons per hectare in Madagascar. The use of bonemeal, rock phosphate, and intercropping with nitrogen-fixing lupin species have significantly contributed to increases in potato yields in Bolivia.[15]

These examples need to be researched to understand why and, importantly, to replicate, improve, and scale up globally. This will close the yield gap and has the potential to overtake the conventional average.

TWO KEY AREAS WHERE ORGANIC HAS HIGH YIELDS

While organic agriculture currently may have lower average yields than the chemically intense industrial agricultural systems in good climate years, there are two areas in which organic agriculture can often have higher yields: under conditions of climate extremes and in traditional smallholder systems. Both of these areas are critical to achieving global food security.

GREATER RESILIENCE IN ADVERSE CONDITIONS

According to research by NASA, the United Nations Framework Convention on Climate Change, and others, the world is seeing increases in the frequency of extreme weather events such as droughts and heavy rainfall. Even if we stopped polluting the planet with greenhouse gases tomorrow, it would take many decades to reverse climate change. Farmers thus have to adapt to the increasing intensity and frequency of adverse and extreme weather events such as droughts and heavy, damaging rainfall.

Published studies show that organic farming systems are more resilient to the emerging weather extremes and can produce higher yields than conventional farming systems in such conditions.[16] For instance, the Wisconsin Integrated Cropping Systems Trials found that organic yields were higher in drought years and the same as conventional in normal weather years.[17]

Similarly, the Rodale Farming Systems Trial (FST) showed that the organic systems produced more corn than the conventional system in drought years. The average corn yields during the drought years were from 28 percent to 34 percent higher in the two organic systems. The yields were 6,938 and 7,235 kilograms per hectare in the organic animal and the organic legume systems respectively, compared with 5,333 kilograms per hectare in the conventional system (Pimentel, 2005). The researchers attributed the higher yields in the dry years to the ability of the soils on organic farms to better absorb rainfall. This absorption is due to the higher levels of organic carbon in those soils, which makes them more friable and better able to store and capture rainwater, which can then be used for crops.[18]

ORGANIC CORN VS. CONVENTIONAL CORN

The corn grown on the organically managed soil (left) in the long-term Rodale Farming Systems Trial has greater drought tolerance than the conventionally grown corn (right) due to better water-holding capacity.

SOIL CLODS

These jars contain the same soil. The soil on the left has higher levels of organic matter due to long-term organic management compared the conventionally managed soil on the right. The conventional soil easily erodes and disperses in the water, whereas the organic soil keeps its integrity and resists erosion.

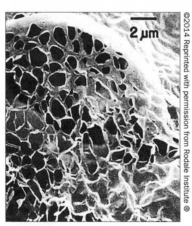

Humus is one of the most important factors of soil organic matter. Its spongelike structure allows it to hold up to thirty times its own weight in water. It is a polymer that glues the soil particles together to give the soil stability, and it holds many of the nutrients that plants need to grow well.

Research also shows that organic systems use water more efficiently due to better soil structure and higher levels of humus and other organic matter compounds. D. W. Lotter and colleagues collected data over ten years during the Rodale FST. Their research showed that the organic manure system and organic legume system (LEG) treatments improve the soils' water-holding capacity, infiltration rate, and water-capture efficiency. The LEG maize soils averaged a 13 percent higher water content than conventional system soils

FiBL DOK TRIALS: WINTER WHEAT UNDER CONVENTIONAL MANAGEMENT

Andreas Fliessbach, FiBL

Heavy rainfall just after planting in the conventionally managed FiBL DOK trials system causes soil to erode and disperse, preventing much of water from infiltrating as well as damaging the new crop and lowering yields.

FiBL DOK TRIALS: WINTER WHEAT UNDER ORGANIC MANAGEMENT

Andreas Fliessbach, FiBL

The same soil in the organically managed system of the trials maintains its structure, resists erosion, and allows the heavy rainfall to infiltrate and to be stored in the soil. Consequently there are higher yields when there are heavy rainfall events at planting.

at the same crop stage, and 7 percent higher than the conventional soils in soybean plots.[19]

The more porous structure of organically treated soil allows rainwater to penetrate more quickly, resulting in less water loss from runoff and higher levels of water capture. This was particularly evident during the two days of torrential downpours from Hurricane Floyd in September 1999, when the organic systems captured around double the water of the conventional systems.[20] Long-term scientific trials conducted by the Research Institute of Organic Agriculture (FiBL) in Switzerland comparing organic, biodynamic, and conventional systems (the DOK trials) had similar results, showing that organic systems were more resistant to erosion and better at capturing water.

This information is significant as the majority of world farming systems are rain fed. The world does not have the resources to irrigate all of the agricultural lands, nor should such a project be started as damming the world's watercourses, pumping from all the underground aquifers, and building millions of kilometers of channels would cause an unprecedented environmental disaster.

Improving the efficiency of rain-fed agricultural systems through organic practices is the most efficient, cost-effective, environmentally sustainable, and practical solution to ensure reliable food production in the increasing weather extremes being caused by climate change.

SMALLHOLDER FARMER YIELDS

The other critical area where research is showing higher yields for good practice organic systems is in traditional smallholder systems. This is very important information as over 85 percent of the world's farmers fall into this category.

A 2008 report by the United National Conference on Trade and Development (UNCTAD) and the United Nations Environment Programme (UNEP) that assessed 114 projects in 24 African countries covering 2 million hectares and 1.9 million farmers found that organic agriculture increases yields in sub-Saharan Africa by 116 percent. There was a 128 percent increase for East Africa. The report notes that despite the introduction of conventional agriculture in Africa, food production per person is 10 percent lower now than in the 1960s. "The evidence presented in this study supports the argument that organic agriculture can be more conducive to food security in Africa than most conventional production systems, and that it is more likely to be sustainable in the long term," stated Supachai Panitchpakdi, secretary general of UNCTAD, and Achim Steiner, executive director of UNEP.[21]

> The study showed that organic farming can yield up to
>
> **THREE TIMES MORE FOOD**
>
> on individual farms in developing countries, as compared to conventional farms.

Badgley et al. from the University of Michigan compared a global dataset of 293 examples of organic versus conventional food production and estimated the average yield ratio. The comparison was divided into different food categories for the developed and the developing world. The researchers found that for most food categories, the average organic yield ratio was slightly less than the average in the developed world and greater than the average in the developing world. Most significantly the study showed that organic farming can yield up to three times more food on individual farms in developing countries, as compared to conventional farms.[22]

This information is especially relevant as Food and Agriculture Organization of the United Nations (FAO) data shows that 80 percent of the food in the developing world comes from smallholder farmers.[23] The developing world is also the region where most of the 850 million undernourished people in the world live, the majority of which are smallholder farmers. With a more than 100 percent increase in food production in these traditional farming systems, organic agriculture provides an ideal solution to end hunger and ensure global food security.

Information published by the ETC Group shows that 70 percent of the world's food is produced by smallholders and only 30 percent by the agribusiness sector.[24] Increasing the yields in the 30 percent of food that comes from the agribusiness sector will show little benefit for two reasons.

Firstly, this sector is already high yielding, and it has very little scope for large increases in yields, such as the more than 100 percent that can be achieved by organic methods in traditional smallholder systems. Secondly, this sector is largely focused on the commodity supply chain. The large food surpluses produced in this sector have not lowered the number of people who are hungry, despite the fact that the world currently produces more than double the amount of food needed to feed everyone. According to FAO figures, the number of hungry has been steadily increasing since 1995. Simply put: the people who need this food the most cannot afford to buy it. On the other hand the people who need it the least are consuming too much, leading to an obesity epidemic around the world. Increasing the production in the agribusiness sector will not solve the current hunger problem, as it cannot do it now, and will most likely increase the obesity epidemic.

INCREASING THE YIELDS IN SMALLHOLDER FARMERS IS THE KEY TO FOOD SECURITY

About 50 percent of the world's hungry are smallholder farmers and 20 percent are the landless poor who rely on smallholders for their employment.[25]

Logically, increasing the yields in the smallholder farmer sector is the key to ending hunger and achieving food security. Organic methods are the most suitable because the necessary methods and inputs can be sourced on farm as well as locally at very little to no cost to the farmers. Conventional systems have largely failed to provide consistently higher yields to the poorest farmers because the expensive synthetic chemical inputs have to be purchased. Most of these farmers do not have the income to do this. It is an inappropriate economic model for the world's most vulnerable farmers, whereas organic agriculture is an appropriate model.

An example of sustainable farming's relevance to smallholder farmers is found in research conducted in the Philippines by

MASIPAG. The yields of organic rice were similar to conventional systems; however a comparison of the income between similar-sized conventional and organic farms found that the average income for organic farms was 23,599 pesos compared 15,643 pesos for the conventional farms. Very significantly, when the household living expenses were deducted from the income the study found that the organic farms had a surplus of 5,967 pesos whereas the conventional farms had a loss of 4,546 pesos at the end of the year, driving them further into debt.[26]

TIGRAY, ETHIOPIA

Another good example is the Tigray Project, managed by the Institute of Sustainable Development (ISD) in Tigray, Ethiopia. This was an area regularly affected by famines that caused many people to die. ISD worked in cooperation with the farmers to revegetate their landscape to restore the local ecology and hydrology. The bio-

AVERAGE MEAN GRAIN YIELDS FOR FOUR CEREALS AND ONE PULSE CROP FROM TIGRAY, NORTHERN ETHIOPIA, 2000–2006 INCLUSIVE

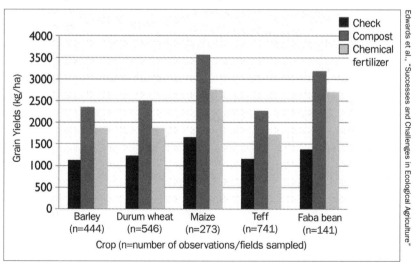

In every case the yield from applying compost was more than 100 percent higher than the traditional systems (Check) and was higher than using chemical fertilizers.

mass from this vegetation was then sustainably harvested to make compost and to feed biogas digesters. This compost was applied to the crop fields. The result after a few years was more than 100 percent increases in yields, better water use efficiency, and greater pest and disease resistance in the crops.

The farmers used the seeds of their own landraces, which had been developed over millennia to be locally adapted to the climate, soils, and the major pests and diseases. The best of these farmer-bred varieties proved very responsive to producing high yields under organic conditions.

The major advantage of this system was that the seeds and the compost were sourced locally at no or little cost to the farmers, whereas the seeds and synthetic chemical inputs in the conventional systems had to be purchased. Not only did the organic system have higher yields, it produced a much better net return to the farmers.[27]

Dr. Sue Edwards, the lead author of the Tigray Project, has produced the following figures on the financial benefits of using compost over chemical fertilizers.

Cost-Benefit Analysis for the Farmer Using Chemical Fertilizer:
- Cost in 2012 was U.S. $300 per hectare for fertilizer (urea + DAP) and pesticides
- Average yield of durum wheat grown with chemical fertilizer 4.5 tons per hectare
- Sold at U.S. $45 per 100 kilograms, farmers' gross income would be U.S. $2,025
- Net income after repaying credit would be U.S. $1,725 per hectare

Cost-Benefit Analysis for the Farmer Using Compost:
- Average rate of compost application is eighty sacks per ha (approximately 8 tons per hectare)
- Opportunity costs for making compost are virtually none as it is all family labor
- Yield of durum wheat grown with compost 6.5 tons per hectare

- Sold at U.S. $45 per 100 kilograms, farmers' income would be U.S. $2,925 per hectare
- All income stays with the farmer as there is no credit

Other Benefits for the Farmer from Using Compost
- Increased resistance to wind and water erosion
- Farmers avoid debt caused by getting chemical fertilizer on credit—now costing U.S. $90 per 100 kilograms
- Farmers making bioslurry compost can sell one sack (approximately 100 kilograms) for U.S. $5.80
- Competent farmers make over thirty-five to one hundred tons of compost a year[28]

This project, using simple, appropriate organic methods, took a region that was previously regularly affected by severe famines harsh enough to kill people through to a food surplus and relative prosperity. The people could now afford to eat well, to buy clothes, send their children to school, pay for medical treatment, afford transport into town, and build adequate houses.

The Tigray Project started in 1996 in four local communities in the central, eastern, and southern parts of the Tigray Regional State and was implemented by the Institute for Sustainable Development (ISD). The Third World Network provided the initial funding. This project is still ongoing with ISD working with the Ethiopian Bureau of Agriculture and Rural Development, *woreda* (district) experts, and development agents to continue executing the Tigray Project. The funding from several donor agencies is assisting in scaling up the project's scope so that more regions in Ethiopia can adopt the practices.[29]

REDUCING THE RISK FROM PESTICIDES BY REPLACING THEM

One of the most effective ways to reduce the health and environmental risks from pesticides is to replace them with non-chemical methods. Organic farming is not a system of neglect. It negates the need for synthetic pesticides by using cultural and ecological management systems as the primary control for pests, weeds, and disease, with a limited use of natural biocides of mineral, plant, and biological origin as the tools of last resort.

The pesticides used in organic systems are from natural sources and are permitted to be used only if they rapidly biodegrade, which means that there are no residues on the products that people consume. By using cultural and ecological methods as the primary management tools with the aims of firstly preventing pests and secondly controlling them, the use of these pesticides is minimal. Research shows that where these natural pesticides are used in organic systems the amounts are over 90 percent less than the synthetic pesticides used in conventional farming.[30]

ECO–FUNCTIONAL INTENSIFICATION

An emerging strategy for replacing pesticides, including natural ones, advocates using ecological management systems that can provide functional services, such as using natural enemies to control pests. The key is to identify these eco-functions and then intensify them in the farming systems so that they replace the need for insecticides. Eco-functional intensification (EFI) is used in organic agriculture to utilize ecological processes rather than chemical intensification. A good example of this is adding insectaries into the farming system. Insectaries are groups of plants that attract and host the beneficial arthropods (insects, bugs, spiders, etc.) and higher animal species. These are the species that eat arthropod pests in farms, orchards, and gardens. They are known collectively as beneficials or natural enemies.

Many beneficial arthropods have a range of host plants. Some useful species—such as parasitic wasps, hoverflies, and lacewings—have carnivorous larvae that eat pests; however the adult stages need nectar and pollen from flowers to become sexually mature and reproduce. Flowers provide beneficial arthropods with concentrated forms of food (pollen and nectar) and increase their chances of surviving, immigrating, and staying in the area. Very importantly, flowers also provide mating sites for beneficials, allowing them to increase in numbers. Without these flowers on a farm the beneficial species die and do not reproduce. Most farming systems eliminate these types of plants as weeds, so consequently they do not have enough beneficials to get effective pest control.

The current loss of biodiversity on this planet is causing the greatest extinction event since the end of the Cretaceous period. Agriculture is one of the main causes due to both habitat loss by clearing forests and the disruption caused by synthetic chemicals. Organic agriculture has a role in conserving and, equally important, increasing and utilizing biodiversity through the concept of eco-functional intensification.

EFI is about utilizing the science of applied agroecology to actively increase the biodiversity in agricultural systems to reduce pests rather than using the conventional approach, based on reductionist monocultures that rely on externally sourced toxic synthetic inputs.

Considering the small average yield difference between chemically intensive and organic farming systems, the more than $50 billion spent annually on conventional farming would be better spent on researching non-chemical, science-based ecological solutions.

PUSH-PULL SYSTEM

The push-pull method in maize (corn) is an excellent example of a science-based eco-functional intensification system that integrates several ecological elements to achieve substantial increases in yields. The possibilities of this method are significant because maize is the key food staple for smallholder farmers in Africa, Latin America, and in many parts of Asia.

Corn stem borers are one of the most significant pests in maize. Conventional agriculture relies on a number of toxic, synthetic pesticides to control these pests. Recently it has started to adopt genetically engineered varieties of corn that produce their own pesticides to combat this pest.

The push-pull system was developed by scientists in Kenya at the International Centre of Insect Physiology and Ecology (ICI-PE); Rothamsted Research, UK; and with the collaboration of other partners.

THE PUSH-PULL SYSTEM

Pull: The Napier grass attracts the moths to lay their eggs in it instead of the maize. Push: The desmodium repels the moths. Desmodium suppresses weeds, especially striga.

Silverleaf desmodium is planted in the crop to repel stem borers and to attract the natural enemies of the pest. The desmodium gives off phenolic compounds that repel the stem borer moth. Its root exudates also stop the growth of many weed species, including striga, which is a serious parasitic weed of maize. Napier grass, a host plant of the moth, is planted outside the field as a trap crop for the stem borer. The desmodium repels (pushes) the pests from the maize and the Napier grass attracts (pulls) the stem borers out of the field to lay their eggs on it in instead of the maize. The sharp silica hairs and sticky exudates on the Napier grass also kill the stem borer larvae when they hatch, breaking the life cycle and reducing pest numbers.

Over forty thousand smallholder farmers in East Africa have adopted this farming system and have seen their maize yields increase from one ton per hectare to three and a half tons. This is a more than 300 percent increase in yields and shows the huge benefits of shifting research away from toxic chemicals to science-based ecological systems.

High yields are not the only benefits. The system does not need synthetic nitrogen because desmodium is a legume and fixes ni-

Desmodium suppresses weeds, adds nitrogen (so there is no need for synthetic nitrogen fertilizers), conserves the soil, repels pests, and provides high-protein stock feed.

The Napier grass is progressively cut and fed to a cow. The excess fresh milk is sold daily as a cash income.

trogen. Soil erosion is prevented due to a permanent ground cover. Very significantly, the system also provides quality fodder for stock.

One farming innovation to improve this system has been to systematically strip harvest the Napier grass to use as fresh fodder for livestock. Livestock can also graze down the field after the maize is harvested. Many push-pull farmers integrate a dairy cow into the system and sell the milk that is surplus to their family needs to provide a regular source of income. This method has provided the farm families with food and income security and has taken them from hunger and desperate poverty into relative prosperity.

There are many examples of other innovative EFI systems that are being developed, such as the system of rice intensification, organic no-/low-till systems (i.e., cover cropping and pasture cropping), agroforestry, and holistic grazing.

THE URGENT NEED FOR MORE RESEARCH

Research into organic agriculture has been chronically underfunded. Trillions of dollars have gone into conventional and GMO research; the organic sector receives a tiny fraction of this. This situation needs to be rectified so that the need for toxic synthetic pesticides is significantly reduced.

Unfortunately in several countries the opposite is occurring. Instead of increasing the investment into organic research, some countries are cutting back on it. Australia discontinued its meager program in 2012, and the U.S. Congress significantly reduced funding in 2013 in its budget cutbacks.

Africa fortunately sees the multiple benefits of organic systems, with the African Union Commission adopting ecological organic agriculture as part of the mix of solutions needed to achieve food security.

Given the small yield difference that has been achieved with trillions of dollars and countless thousands of researchers compared to what organic farmers have achieved when left largely to their own devices, it would have to be argued that the substantial proportion of the funding into conventional agriculture has been a very poor use of valuable research funds. Also given that the new research into organic systems is starting to show very impressive increases in yields, it is logical to argue that research into organic agriculture is a far better use of these research funds.

NOTES

[1] APVMA, "About the APVMA: Factsheet."

[2] Verena Seufert, Navin Ramankutty, and Jonathan A. Foley, "Comparing the Yields of Organic and Conventional Agriculture," *Nature* 485 (May 2012): 229–32, http://www.nature.com/nature/journal/v485/n7397/full/nature11069.html; Tomek de Ponti, Bert Rijk, and Martin van Ittersum, "The Crop Yield Gap between Organic and Conventional Agriculture," *Agricultural Systems* 108 (2012): 1–9.

[3] Catherine Badgley et al., "Organic Agriculture and the Global Food Supply," *Renewable Agriculture and Food Systems* 22, no. 2 (2007): 86–108.

[4] Urs Niggli, "Sustainability of Organic Food Production: Challenges and Innovations," *Proceedings of the Nutrition Society*, forthcoming.

[5] Alfredo Flores, "Organic Pecans: Another Option for Growers," *Agricultural Research* magazine, U.S. Agricultural Research Service, November/December 2008, http://www.ars.usda.gov/is/AR/archive/nov08/pecans1108.pdf.

[6] Jean-Paul Chavas, Joshua L. Posner, and Janet L. Hedtcke, "Organic and Conventional Production Systems in the Wisconsin Integrated Cropping Systems Trial: II. Economic and Risk Analysis 1993–2006," *Agronomy Journal* 101, no. 2 (2009): 253–60, http://wicst.wisc.edu/wp-content/uploads/pages-from-wicst-econ-aj2009.pdf.

[7] David Pimentel et al., "Environmental, Energetic and Economic Comparisons of Organic and Conventional Farming Systems," *Bioscience* 55, no. 7 (July 2005): 573–82, http://www.ce.cmu.edu/~gdrg/readings/2007/02/20/Pimental_EnvironmentalEnergeticAndEconomicComparisonsOfOrganicAndConventionalFarmingSystems.pdf.

[8] Rodale Institute, "Organic No-Till," http://www.rodaleinstitute.org/no-till_revolution (accessed January 2014).

[9] Bob Turnbull, "Research Shows Organic Corn, Soybean Yields Can Exceed Conventional," *Organic & Non-GMO Report*, January 2010, http://www.non-gmoreport.com/articles/feb10/organic_corn_soybean_yields_exceed_conventional.php.

[10] Lorenz Bachman, Elizabeth Cruzada, and Sarah Wright, "Food Security and Farmer Empowerment: A Study of the Impacts of Farmer-Led Sustainable Agriculture in the Philippines," MASIPAG, Anos Los Banos, Laguna 4000, Philippines, 2009.

[11] George Monbiot, "Organic Farming Will Feed the World," *Guardian*, August 24, 2000.

[12] John P. Reganold et al., "Sustainability of Three Apple Production Systems," *Nature* 410 (April 2001): 926–30.

[13] Laurie E. Drinkwater, Peggy Wagoner, and Marianne Sarrantonio, "Legume-Based Cropping Systems Have Reduced Carbon and Nitrogen Losses," *Nature* 396 (November 1998): 262–65, http://www.nature.com/nature/journal/v396/n6708/abs/396262a0.html.

[14] Rick Welsh, "The Economics of Organic Grain and Soybean Production in the Midwestern United States," Policy Studies Report No. 13, Henry A. Wallace Institute for Sustainable Agriculture, May 1999.

[15] Nicholas Parrott and Terry Marsden, *The Real Green Revolution: Organic and Agroecological Farming in the South* (Canonbury Villas, London: Greenpeace Environmental Trust, 2002).

[16] Pimentel et al., "Environmental, Energetic and Economic Comparisons of Organic and Conventional Farming Systems"; Drinkwater, Wagoner, and Sarrantonio, "Legume-Based Cropping Systems Have Reduced Carbon and Nitrogen Losses"; Welsh, "The Economics of Organic Grain and Soybean Production."

[17] Chavas, Posner, and Hedtcke, "Organic and Conventional Production Systems in the Wisconsin Integrated Cropping Systems."

[18] Tim J. LaSalle and Paul Hepperly, "Regenerative Organic Farming: A Solution to Global Warming," Rodale Institute, 2008.

[19] D. W. Lotter, R. Seidel, and W. Liebhart, "The Performance of Organic and Conventional Cropping Systems in an Extreme Climate Year," *American Journal of Alternative Agriculture* 18, no. 3 (2003): 146–54, http://www.donlotter.net/lotter_ajaa_article.pdf.

[20] Ibid.

[21] Rachel Hine and Jules Pretty, *Organic Agriculture and Food Security in Africa*, United Nations Environment Programme, United Nations Conference on Trade and Development, September 2008, http://www.unctad.org/en/docs/ditcted200715_en.pdf.

[22] Badgley et al., "Organic Agriculture and the Global Food Supply."

[23] Food and Agriculture Organization of the United Nations, "The Challenge," chapter 1 in *Save and Grow*, Rome, 2011, http://www.fao.org/ag/save-and-grow/en/1/index.html.

[24] ETC Group, "Who Will Feed Us? Questions for the Food and Climate Crises," November 1, 2009, http://www.etcgroup.org/content/who-will-feed-us.

[25] Ibid.

[26] Bachman, Cruzada, and Wright, "Food Security and Farmer Empowerment."

[27] Sue Edwards, Tewolde Berhan Gebre Egziabher, and Hailu Araya, "Successes and Challenges in Ecological Agriculture: In Experiences from Tigray, Ethiopia," in *Climate Change and Food Systems Resilience in Sub-Saharan Africa*, ed. Lim Li Ching, Sue Edwards, and Nadia El-Hage Scialabba (Rome: Food and Agriculture Organization of the United Nations, 2011), 231–94. Available online at www.fao.org/docrep/014/i2230e/i2230e09.pdf.

[28] Dr. Sue Edwards, personal communication.

29 Ibid.

30 Paul Maeder et al., "Soil Fertility and Biodiversity in Organic Farming," Science 296 (May 2002): 1694–97, http://www.sciencemag.org/content/296/5573/1694.short.

CONCLUSION

The Romans lined their drinking vessels with lead, and the ancient Chinese fed their emperors immortality pills made from mercury. At that time these cultures did not know about the slow-developing fatal illnesses caused by regular small exposures to these poisons. Unlike the Romans and the ancient Chinese, as a society we do know that the chemicals used in our food production are toxic because we've seen their ability to kill pests, diseases, and weeds.

History will look back with amazement that not only did our regulatory authorities know that these substances were toxic, for decades they ignored a huge body of hundreds of credible scientific studies written by several hundred scientists that documented the multiple harmful effects they cause to humans and the environment.

Future historians will debate the reasons as to why a technologically advanced civilization would allow such a situation to develop. Unlike the Romans and ancient Chinese, our inaction cannot be attributed to ignorance because the extensive body of information on the multiple health problems caused by these substances is openly available in published literature produced by some of the most intelligent sections of society: scientists and researchers.

They will ponder whether the cause of this inaction was incompetence, laziness, corruption, protecting reputations that were built on outdated scientific methodologies, or greed on the part of a few to generate great wealth at the expense of many. Possibly they will conclude that all of these factors had a role.

THE UNBORN AND GROWING CHILDREN

The issue that will puzzle these future historians the most is how and why such an advanced civilization could permit these toxic compounds to damage their children, causing numerous health problems in future generations. Some of the most important issues documented throughout this book are the special needs of the unborn and growing children. This group is the most vulnerable to the harm caused by chemicals. The research shows that they are being exposed to cocktails of chemicals even before they are born. As young children they have the highest levels of pesticide exposure due to their food consumption in relation to their body weight. Of particular concern is that the fetus and newborn possess lower concentrations of protective serum proteins than adults. A major consequence of this vulnerability is a greater susceptibility to cancers and developmental neurotoxicity, where the poison damages the developing nervous system.

They are more vulnerable than adults to the effects of endocrine disrupters because their tissues and organs are still developing and rely on balanced hormone signals to ensure that they develop in orderly sequences. Small disruptions in these hormone signals by endocrine-disrupting chemicals can significantly alter the way these body parts and metabolic systems develop. These altered effects will not only last a lifetime; they can be passed on to future generations.

A large body of published, peer-reviewed scientific research shows that pesticide exposure in unborn and growing children is linked to:

- Cancers
- Thyroid disorders
- Immune system problems
- Lower IQs
- Attention deficit hyperactivity disorder
- Autism spectrum disorders
- Lack of physical coordination
- Loss of temper—anger management issues
- Bipolar/schizophrenia spectrum of illnesses
- Depression
- Digestive system problems
- Cardiovascular disease
- Reproductive problems (as adults)
- Deformities of the genital-urinary systems
- Changes to metabolic systems, including childhood obesity and diabetes

The current pesticide-testing methodologies use adolescent and adult animals. Consequently, they will not detect adverse health issues that are specific to the unborn and children. The U.S. EPA's approach of lowering residues by a factor of ten for children is based on data-free assumptions, especially since the evidence coming from endocrine disruption and non-monotonic doses shows that in many cases the exposure should be more than a thousand times lower.

THE USE OF PEER-REVIEWED SCIENTIFIC STUDIES PUBLISHED IN CREDIBLE JOURNALS

There is a critical need for all regulatory decisions to be made on the basis of credible scientific evidence, largely based on peer-reviewed studies published in credible journals. The interpretation of scientific data is not always clear-cut because of many variables, especially where there are gaps in the data. Publishing studies in journals so that they are available for all the relevant stakeholders to read and analyze allows for a wider and more critical debate over the data and encourages a more rigorous process in reaching conclusions.

One of the most important aspects of this scientific process is that studies should clearly document the materials and methods used in research experiments. Many studies showing the adverse effects of pesticides and problems with current regulatory methodologies are criticized through academic and political debates. The accepted way to resolve the credibility of research is to accurately repeat the experiment by using the material and methods described in the published paper to see if they consistently produce the same results. This will in most cases confirm whether or not the research conclusions are correct.

One of the methods used at times by the pesticide industry and regulators to rebut studies is to state that industry studies fail to report the same adverse outcomes. These industry studies are largely unpublished and are usually based on different criteria than the peer-reviewed studies they are meant to refute. Setting up research using different criteria will most likely result in different outcomes. When the outcomes of these "similar" studies do not confirm the results of the study with adverse health outcomes, the pesticide industry and regulators use them to discredit the potentially profit-damaging study and dismiss its results.

A good analogy would be if an organization performed a few studies on a select number of elderly people who smoke tobacco and then announced that "studies" show no evidence that smoking reduces a person's life span, as these people have lived long lives while smoking every day. Tobacco proponents could then use these biased studies as "evidence" that the hundreds of other studies linking smoking tobacco to numerous health issues should be ignored because the results are "not proven."

An example was given by Dan Fagin in a comment article that he wrote in *Nature* in 2012, a few months after of the publication of the comprehensive meta review on endocrine disruption by Vandenberg et al. He mentions two separate studies that were conducted on the plasticizer bisphenol A (BPA) to assess if it is an endocrine disrupter. One study was conducted by the U.S. Food and Drug Administration (FDA) and the other by a private firm that was contracted by industry. Neither study found evidence of endocrine disruption by BPA, despite numerous other studies finding this.[1] Vandenberg et al., for example, reported that "In 2006, vom Saal and Welshons ... examined the low-dose BPA literature, identifying more than 100 studies published as of July 2005 that reported significant effects of BPA below the established LOAEL, of which 40 studies reported adverse effects below the 50μg/kg·d safe dose set by the EPA and U.S. Food and Drug Administration (FDA)."[2]

Substantially more studies have been published since 2005 showing the endocrine-disrupting effects of BPA. Because regulatory authorities take a reactionary approach rather than a precautionary approach, one study confirming the status quo tends to take precedence over the many studies that challenge it.

According to Fagin, largely because of these two studies, neither the U.S. FDA nor the U.S. EPA will alter their risk assessments for BPA despite more than a hundred published, peer-reviewed scientific studies showing adverse health effects. "The FDA still says that BPA has no adverse effects at levels below 50 milligrams per kilogram of body weight per day—a level that vom Saal contends should actually be two million times lower, at 25 nanograms."[3]

The vom Saal quoted by Fagin is Dr. Frederick vom Saal, PhD, a neurobiologist and professor at the University of Missouri–Columbia. He is a leading and pioneering scientist in the field of endocrine disrupters and has been since the 1970s. He was one of the coauthors of the Vandenberg et al. meta review. Vom Saal stated that both of the studies conducted by the industry and the FDA used criteria that were not suitable for finding the effects of endocrine disruption. Parts of the design of the two experiments were regarded by vom Saal and other expert researchers in the field of endocrine disruption as insensitive to low-dose effects, and consequently they would not be found in the results.

Vandenberg et al. give many examples of the way the differences in the design of experiments will result in outcomes that will not confirm the earlier studies. "In fact, the NTP [National Toxicology Program] low-dose panel itself suggested that factors such as strain differences, diet, caging and housing conditions, and seasonal variation can affect the ability to detect low dose effects in controlled studies."[4]

A review of the studies that were used to refute the toxic effects of low doses of atrazine found many flaws in the design of the experiments. "Hayes' work also clearly addressed the so called irreproducibility of these findings by analyzing the studies that were unable to find effects of the pesticide; he noted that the negative studies had multiple experimental flaws, including contamination of the controls with atrazine, overcrowding (and therefore underdosing) of experimental animals, and other problems with animal husbandry that led to mortality rates above 80%."[5]

These examples highlight the need to accurately replicate the material and methods used in the original studies when testing whether their results are credible rather than designing similar studies using variations of the materials and methods. In reality these "similar" studies are entirely new studies because they generally use a different set of criteria. Researchers are not replicating the original study, and therefore it should be expected that they will see different results than the original study.

Similarly, just because one study does not find an adverse health outcome like cancer or endocrine disruption, it does not necessarily invalidate a study that does. It usually means that the studies are using different criteria and methodologies. Glyphosate is a good example: several animal feeding studies did not find any evidence of cancer, but there is a study linking glyphosate to non-Hodgkin's lymphoma. There are several studies linking it and its formulations to gene mutations, cell-cycle dysregulation, and chromosomal aberrations. These types of genetic damage can be precursors to cancer. A study of human breast cancer cells found that glyphosate caused a rapid multiplication of the cancer cells. Instead of dismissing the studies, regulatory authorities need to investigate the criteria and methodologies used in these studies in order to fully understand why the cancer growths and pre-cancer events occurred.

The strategy of using studies that do not find adverse health problems to cast doubt on the credibility of a study that has is a tactic used by some industries and regulators. Big tobacco, the lead industry, and the asbestos industry did this for decades before public pressure working in partnership with concerned scientists finally forced the governments and regulatory authorities to implement some of the necessary changes. This inaction has resulted in and continues to cause millions of people to suffer from painful and needless illnesses and early deaths. Most recently this technique was used to sow the seeds of doubt about the science of human-created greenhouse gases as the major cause of climate change as well as junk food composed of empty calories as a cause of the global obesity epidemic.

The pesticide industry has a long history of muddying the waters with false comparisons like this and has been very successful at convincing most people to believe the myths that their food is safe.

THE NEED FOR CHANGES IN THE METHODOLOGIES USED TO TEST CHEMICALS

One of the major issues repeated consistently in this book is the need for changes in the current approaches and methodologies used by regulatory authorities in assessing the safety of pesticides.

The huge body of missing information needs to be researched, and the outdated testing methods need to be augmented with the emerging body of scientific techniques so that they can provide the missing data.

Additional testing needs to be done for:
- Mixtures and cocktails of chemicals
- The actual formulated products, not just the active ingredient
- The toxicity of pesticide metabolites
- The special requirements of fetuses, newborns, and growing children
- Endocrine disruption
- Metabolic disruption
- Intergenerational effects on all organs and physiological systems
- Developmental neurotoxicity

Until this is done, regulatory bodies have no credible scientific evidence backing a statement that any level of pesticide residue is safe for humans or the environment.

DATA-FREE ASSUMPTIONS AS THE BASIS OF PESTICIDE REGULATION

The scientific credibility of pesticide regulatory authorities has to be seriously questioned when they are approving the use of pesticides on the basis of data-free assumptions.

A good example of this is the approval of formulated pesticide products as safe on the basis of just testing one of the ingredients without testing the whole formulation. Given that the other chemical ingredients are chemically active as they are added to the formulations to make the active ingredient work more effectively, the assumption that they are inert and will not increase the toxicity of the whole formulation lacks scientific credibility. There are no requirements to test the toxicity of the whole formulation to generate credible evidence based scientific data. This means that the current approval process is based on the data-free assumption that the "inerts" do not alter the toxicity of the active ingredient.

Regulatory authorities approve several different pesticides for a crop—such as herbicides, fungicides, and insecticides—on the basis that all of them can be used in the normal production of the crop. Consequently, multiple residues will be found in the crop; residue testing found that 47.4 percent of food in the United States had two or more pesticide residues. The current approval process of testing each pesticide separately is based on the assumption that if each chemical is safe individually then the combinations of these chemicals are also safe. There are a number of published scientific studies showing that combinations of pesticide residues can cause serious adverse health outcomes due to additive or synergistic effects. The failure to test the combinations of approved pesticides for potential health risks means that regulatory authorities do not have any evidence-based data indicating that these residue combinations are safe. The current data-free assumption of safety used by regulatory authorities lacks scientific credibility.

The lack of testing for the metabolites of pesticides, given that limited testing shows that many of them are more toxic and residual than the pesticide itself, is another massive data gap. Once again, approval has been based on data-free assumptions of safety.

The setting of the ADI is another example. Given that there are hundreds of studies showing that many chemicals can be endocrine disruptors and therefore more toxic at lower doses, setting the ADI on the basis of extrapolating it from testing done at higher doses is another data-free assumption. The only way to ensure that the ADI is safe and does not act as an endocrine disruptor is to do the testing at the actual residue levels that are set for the ADI.

The special requirements of the fetus, the newborn, and the growing child in relation to developmental neurotoxicity are also subject to data-free assumptions. Currently the pesticide testing used in the regulatory approval processes does not specifically test for any of the risks particular to these age groups, and the ADIs are set based on the testing of adolescent animals. Until testing is specifically designed to assess the dangers to the developing fetus and the very young, there is no evidence-based data specific to this age group. Once again, pesticide ADIs are approved as safe for children without credible scientific evidence to prove their safety.

It is the same with intergenerational effects. Unless testing is done over several generations, especially on organs and physiological processes, these is no data to show that the current ADIs will not cause health problems for the future generations. There are many scientific studies showing that exposure to pesticide residues cause adverse health problems in future generations, so ignoring this issue could prove dangerous. It is a data-free assumption to approve pesticides on the basis that these intergenerational effects are not a significant issue.

The regulation of pesticides should be based on data generated through credible scientific studies and testing, not on data-free assumptions as it is currently.

THE POSITIVE ALTERNATIVE AND FUTURE

Even if regulatory authorities started tomorrow, it would take decades and billions of dollars in funds to test all the registered pesticide products and the thousands of common combinations to acquire the relevant missing data needed to establish the safe use of these poisons.

In the meantime a precautionary approach to replace pesticides is the best strategy because reducing pesticide exposure to lower levels gives no guarantee of safety. Currently, due to the numerous significant data gaps, there is no credible science to show that any level of residue is safe. Adopting farming systems that replace pesticides with nontoxic, natural methods of pest control is the most effective and logical way to avoid the current uncertainties surrounding pesticide use.

Research has clearly shown that organic agriculture can get the yields needed to feed the poor and the hungry, especially in the case of smallholder agriculture—the majority of the world's farmers.

It is critically important that a substantial proportion of the billions of dollars spent on research and development of chemically intensive agriculture are invested on researching the possibilities of the emerging high-yielding organic systems. Using research and development to replicate, improve, and scale these systems up globally will enable agriculture to achieve high yields without the use of toxic chemicals.

REDUCED PESTICIDE USE MAKES
NO DIFFERENCE IN SAFETY

There are many certified food-labeling systems that portray themselves as ecological or sustainable because of reduced pesticide use and perpetuate the myth of "safe food." In the case of "good agricultural practices," all pesticides permitted by regulators can be used as long as they are used according to the label on the container. The assumption is that as long as the pesticides are used per the label's instructions, they are safe.

Some of these systems use the WHO's toxicity classification as the basis of safe use. They prohibit the use of the most acutely toxic chemicals based on the LD_{50}s, but they allow the use of thousands of other toxic pesticide formulations. As stated in chapter 1, LD_{50}s are used to determine the acute toxicity of a chemical (the toxicity that will quickly cause death) but are irrelevant in showing the longer-term toxic effects of a chemical or formulated mixtures, such as cancers, cell mutations, endocrine disruption, birth defects, organ and tissue damage, nervous system damage, behavior changes, and immune system damage.

Other "safe-food" and "eco-label" systems just prohibit pesticides banned by the European Union (EU) and/or the United States; however as has been shown repeatedly throughout this book, the vast majority of the thousands of chemicals used in the EU and the United States have not been tested for safety. This is especially the case with the thousands of commercial pesticide formulations composed of the active ingredient and the "inerts" that have no testing for the numerous adverse health effects that peer-reviewed scientific papers have linked to pesticides. Consumers should therefore be greatly concerned that thousands of these formulations are permitted in "safe-food," "good-agricultural-practice," "sustainable," and "eco-label" certification systems.

These systems cannot give any guarantee that their pesticide use is any safer than conventional systems while they permit the use of any level of synthetic chemical pesticides. Given that the science shows that for many chemicals even the smallest amounts can have serious adverse health effects, especially on the developing fetus and growing children, any residues are potentially unsafe, no matter how small. In these situations, reduced amounts make no difference, and in the case of non-monotonic doses they could even be doing more harm.

HEALTH MUST COME FIRST

Out of all the criteria being used to assess the environmental sustainability of agricultural systems, the health of people and all the biodiverse forms of life in our planet's ecosystems must be our number one priority.

What are the major benefits of having good recycling outcomes, low-carbon footprints, low energy use, better water-use efficiency, locally grown produce, natural, etc., if the production system is severely harming the health of the surrounding environment and the people who consume the products from it? It is an even greater concern when the genetic damage caused by pesticides propagates a harmful legacy that will be passed onto future generations.

Is it better to have the freshness of locally grown produce that is toxic or a nontoxic product that may not be as fresh? Ideally it is best to have fresh, locally grown, and nontoxic, but when the ideal isn't available it is always better to have nontoxic as the first choice. People do not get serious illnesses because of the difference in the distance a product has traveled to get to market. On the other hand there can be serious consequences from even minute chemical residues in the food consumed by mothers being passed through the placenta to the fetus or through breast milk to the newborn, even if it is locally produced.

It is the same with all the other ecological options. Ideally we want good environmental outcomes across all criteria; however when the ideal isn't possible, the health of our future generations must come first. How can we pass on a better world to them when we are passing on generations of adverse health outcomes?

PROTECTING OUR FUTURE AND OUR CHILDREN

Currently, for consumers the only way to avoid these poisons is to eat organically grown food that has been produced with organic guarantee systems such as third-party-certification, participatory guarantee systems (PGS), as a member of an organic consumer supported agriculture (CSA) scheme, or farmers markets that check their farmers' production claims. These guarantee systems will ensure that the food is produced without toxic compounds. Most importantly, scientific studies show that eating organic food results in lower levels of these pervasive chemicals in humans, particularly children.

A study published in *Environmental Health Perspectives* found that children who eat organic fruits, vegetables, and juices can significantly lower the levels of organophosphate pesticides in their bodies. The University of Washington researchers who conducted the study concluded, "The dose estimates suggest that consumption of organic fruits, vegetables, and juice can reduce children's exposure levels from above to below the U.S. Environmental Protection Agency's current guidelines, thereby shifting exposures from a range of uncertain risk to a range of negligible risk. Consumption of organic produce appears to provide a relatively simple way for parents to reduce their children's exposure to OP [organophosphate] pesticides."[6]

Researchers in a 2006 study found that the urinary concentrations of the specific metabolites for malathion and chlorpyrifos decreased to undetectable levels immediately after the introduction of organic diets and remained undetectable until the conventional diets were reintroduced. The researchers from Emory University, Atlanta, Georgia; the University of Washington, Seattle, Washington; and the Centers for Disease Control and Prevention, Atlanta, Georgia, stated, "In conclusion, we were able to demonstrate that an organic diet provides a dramatic and immediate protective effect against exposures to organophosphorus pesticides that are commonly used in agricultural production. We also concluded that these children were most likely exposed to these organophosphorus pesticides exclusively through their diet."[7]

It is time to dispense with the myth that foods from farming systems that use synthetic pesticides are safe to eat. This includes low- or reduced-pesticide farming systems, as there is no credible science to guarantee that any level of exposure is safe. The lack of rigorous testing and the blatant disregard of the current science by regulators means that, until these data gaps are filled, the most logical option is to avoid food from farming systems that use these toxic compounds.

NOTES

[1] Dan Fagin, "Toxicology: The Learning Curve," *Nature* 490, no. 7421 (October 2012): 462–65.
[2] Vandenberg et al., "Hormones and Endocrine-Disrupting Chemicals."
[3] Fagin, "Toxicology."

[4] Vandenberg et al., "Hormones and Endocrine-Disrupting Chemicals."

[5] Ibid.

[6] Cynthia Curl, Richard A. Fenske, and Kai Elgethun, "Organophosphorus Pesticide Exposure of Urban and Suburban Preschool Children with Organic and Conventional Diets," *Environmental Health Perspectives* 111, no. 3 (March 2003): 377–82.

[7] Chensheng Lu et al., "Organic Diets Significantly Lower Children's Dietary Exposure to Organophosphorus Pesticides," *Environmental Health Perspectives* 114, no. 2 (February 2006): 260–63.

SELECTED BIBLIOGRAPHY

Aldridge, Justin, Frederic Seidler, Armando Meyer, Indiro Thillai, and Theodore Slotkin. "Serotonergic Systems Targeted by Developmental Exposure to Chlorpyrifos: Effects during Different Critical Periods." *Environmental Health Perspectives* 111, no. 14 (November 2003): 1736–43.

Antoniou, Michael, Mohamed Ezz El-Din Mostafa Habib, C. Vyvyan Howard, Richard C. Jennings, Carlo Leifert, Rubens Onofre Nodari, Claire Robinson, and John Fagan. "Teratogenic Effects of Glyphosate-Based Herbicides: Divergence of Regulatory Decisions from Scientific Evidence." *Journal of Environmental and Analytical Toxicology* (2012): S4:006. doi:10.4172/2161-0525.S4-006.

Bachman, Lorenz, Elizabeth Cruzada, and Sarah Wright. "Food Security and Farmer Empowerment: A Study of the Impacts of Farmer-Led Sustainable Agriculture in the Philippines." MASIPAG. Anos Los Banos, Laguna 4000, Philippines, 2009.

Badgley, Catherine, Jeremy Moghtader, Eileen Quintero, Emily Zakem, M. Jahi Chappell, Katia Avilés-Vázquez, Andrea Samulon, and Ivette Perfecto. "Organic Agriculture and the Global Food Supply." *Renewable Agriculture and Food Systems* 22, no. 2 (2007): 86–108.

Bellinger, David C. "A Strategy for Comparing the Contributions of Environmental Chemicals and Other Risk Factors to Neurodevelopment of Children." *Environmental Health Perspectives* 120, no. 4 (2012): 501–7, http://dx.doi.org/10.1289/ehp.1104170.

Benbrook, Charles. "Impacts of Genetically Engineered Crops on Pesticide Use in the U.S.—The First Sixteen Years." *Environmental Sciences Europe* 24, no. 24 (September 2012): doi:10.1186/2190-4715-24-24.

Bergman, Åke, Jerrold J. Heindel, Susan Jobling, Karen A. Kidd, and R. Thomas Zoeller, eds. *State of the Science of Endocrine Disrupting Chemicals 2012.* United Nations Environment Programme and the World Health Organization, 2013.

Bouchard, Maryse F., Jonathan Chevrier, Kim G. Harley, Katherine Kogut, Michelle Vedar, Norma Calderon, Celina Trujillo, Caroline Johnson, Asa Bradman, Dana Boyd Barr, and Brenda Eskenaz. "Prenatal Exposure to Organophosphate Pesticides and IQ in 7-Year Old Children." *Environmental Health Perspectives* 119, no. 8 (August 2011): 1189–95. Published online April 21, 2011, http://www.ncbi.nlm.nih.gov/pmc/articles/PMC3237357/.

Buznikov, Gennady A., Lyudmila A. Nikitina, Vladimir V. Bezuglov, Jean M. Lauder, S. Padilla, and Theodore A. Slotkin. "An Invertebrate Model of the Developmental Neurotoxicity of Insecticides: Effects of Chlorpyrifos and Dieldrin in Sea Urchin Embryos and Larvae." *Environmental Health Perspectives* 109, no. 7 (July 2001): 651–61.

Cabello, Gertrudis, Mario Valenzuela, Arnoldo Vilaxa, Viviana Durán, Isolde Rudolph, Nicolas Hrepic, and Gloria Calaf. "A Rat Mammary Tumor Model Induced by the Organophosphorous Pesticides Parathion and Malathion, Possibly through Acetylcholinesterase Inhibition." *Environmental Health Perspectives* 109, no. 5 (May 2001): 471–79.

Cadbury, Deborah. *The Feminization of Nature: Our Future at Risk.* Middlesex, England: Penguin Books, 1998.

Carson, Rachel. *Silent Spring.* New York: Penguin Books, 1962.

Cattani, Daiane, Vera Lúcia de Liz Oliveira Cavalli, Carla Elise Heinz Rieg, Juliana Tonietto Domingues, Tharine Dal-Cim, Carla Inês Tasca, Fátima Regina Mena Barreto Silva, Ariane Zamoner. "Mechanisms Underlying the Neurotoxicity Induced by Glyphosate-Based Herbicide in Immature Rat Hippocampus: Involvement of Glutamate Excitotoxicity." *Toxicology* 320 (March 2014): 34–45.

Cavalli, Vera Lúcia de Liz Oliveira, Daiane Cattani, Carla Elise Heinz Rieg, Paula Pierozan, Leila Zanatta, Eduardo Benedetti Parisotto, Danilo Wilhelm Filho, Fátima Regina Mena Barreto Silva, Regina Pessoa-Pureur, and Ariane Zamoner. "Roundup Disrupted Male Reproductive Functions by Triggering Calcium-Mediated Cell Death in Rat Testis and Sertoli Cells." *Free Radical Biology & Medicine* 65 (December 2013): 335–46.

Cavieres, María Fernanda, James Jaeger, and Warren Porter. "Developmental Toxicity of a Commercial Herbicide Mixture in Mice: I. Effects on Embryo Implantation and Litter Size." *Environmental Health Perspectives* 110, no. 11 (November 2002): 1081–85.

Charizopoulos, Emmanouil, and Euphemia Papadopoulou-Mourkidou. "Occurrence of Pesticides in Rain of the Axios River Basin, Greece." *Environmental Science & Technology* 33, no. 14 (July 1999): 2363–68.

Chavas, Jean-Paul, Joshua L. Posner, and Janet L. Hedtcke. "Organic and Conventional Production Systems in the Wisconsin Integrated Cropping Systems Trial: II. Economic and Risk Analysis 1993–2006." *Agronomy Journal* 101, no. 2 (2009): 253–60, http://wicst.wisc.edu/wp-content/uploads/pages-from-wicst-econ-aj2009.pdf.

Clements, Chris, Steven Ralph, and Michael Petras. "Genotoxicity of Select Herbicides in *Rana catesbeiana* Tadpoles Using the Alkaline Single-Cell Gel DNA Electrophoresis (Comet) Assay." *Environmental and Molecular Mutagenesis* 29, no. 3 (1997): 277–88.

Colborn, Theo. "A Case for Revisiting the Safety of Pesticides: A Closer Look at Neurodevelopment." *Environmental Health Perspectives* 114, no. 1 (January 2006): 10–17.

Colborn, Theo, Dianne Dumanoski, and John Peterson Myers. *Our Stolen Future: Are We Threatening Our Fertility, Intelligence, and Survival? A Scientific Detective Story.* New York: Dutton, 1996.

Cox, Caroline. "Atrazine: Environmental Contamination and Ecological Effects." Northwest Coalition against Pesticides, Eugene, Oregon. *Journal of Pesticide Reform* 21, no. 3 (Fall 2001): 12.

———. "Glyphosate (Roundup)." Northwest Coalition against Pesticides, Eugene, Oregon. *Journal of Pesticide Reform* 24, no. 4 (Winter 2004).

Curl, Cynthia, Richard A. Fenske, and Kai Elgethun. "Organophosphorus Pesticide Exposure of Urban and Suburban Preschool Children with Organic and Conventional Diets." *Environmental Health Perspectives* 111, no. 3 (March 2003): 377–82.

Dallegrave, Eliane, Fabiana DiGiorgio Mantese, Ricardo Soares Coelho, Janaína Drawans Pereira, Paulo Roberto Dalsenter, Augusto Langeloh. "The Teratogenic Potential of the Herbicide Glyphosate-Roundup in Wistar Rats." *Toxicology Letters* 142, nos. 1–2 (April 2003): 45–52.

de Ponti, Tomek, Bert Rijk, and Martin van Ittersum. "The Crop Yield Gap between Organic and Conventional Agriculture." *Agricultural Systems* 108 (2012): 1–9.

Drinkwater, Laurie E., Peggy Wagoner, and Marianne Sarrantonio. "Legume-Based Cropping Systems Have Reduced Carbon and Nitrogen Losses." *Nature* 396 (November 1998): 262–65, http://www.nature.com/ nature/journal/v396/n6708/abs/396262a0.html.

Edwards, Sue, Tewolde Berhan Gebre Egziabher, and Hailu Araya. "Successes and Challenges in Ecological Agriculture: In Experiences from Tigray, Ethiopia." In *Climate Change and Food Systems Resilience in Sub-Saharan Africa*, edited by Lim Li Ching, Sue Edwards, and Nadia El-Hage Scialabba, 231–94. Rome, Italy: Food and Agriculture Organization of the United Nations, 2011. Available online at www.fao.org/docrep/014/i2230e/ i2230e09.pdf.

El-Shenawy, Nahla S. "Oxidative Stress Responses of Rats Exposed to Roundup and Its Active Ingredient Glyphosate." *Environmental Toxicology and Pharmacology* 28, no. 3 (November 2009): 379–85.

Engel, Stephanie M., James Wetmur, Jia Chen, Chenbo Zhu, Dana Boyd Barr, Richard L. Canfield, and Mary S. Wolff. "Prenatal Exposure to Organophosphates, Paraoxonase 1, and Cognitive Development in Children." *Environmental Health Perspectives* 119 (2011): 1182–88. Published online April 21, 2011, http://ehp.niehs.nih.gov/1003183/.

ETC Group. "Who Will Feed Us? Questions for the Food and Climate Crises." November 1, 2009, http://www.etcgroup.org/content/who-will -feed-us.

European Commission. "What is REACH?" January 24, 2014, http:// ec.europa.eu/environment/chemicals/reach/reach_en.htm.

European Food Safety Authority. "Conclusion on the Peer Review of the Pesticide Risk Assessment for Bees for the Active Substance Imidacloprid." *EFSA Journal* 11, no 1 (2013): 3068. Available online at http://www.efsa .europa.eu/en/efsajournal/pub/3068.htm.

EXTOXNET: The Extension Toxicology Network. University of California-Davis, Oregon State University, Michigan State University, Cornell University, and the University of Idaho, 1996, http://extoxnet.orst.edu/. Primary files are maintained and archived at Oregon State University. Accessed online January 24, 2014.

Fagin, Dan. "Toxicology: The Learning Curve." *Nature* 490, no. 7421 (October 2012): 462–65.

Flores, Alfredo. "Organic Pecans: Another Option for Growers." *Agricultural Research* magazine. U.S. Agricultural Research Service, November/December 2008, http://www.ars.usda.gov/is/AR/archive/nov08/pecans1108.pdf.

Food and Agriculture Organization of the United Nations. "The Challenge." Chapter 1 in *Save and Grow*. Rome, 2011, available online at http://www.fao.org/ag/save-and-grow/en/1/index.html.

Food Standards Australia and New Zealand. "20th Australian Total Diet Survey," 2002, available online at http://www.foodstandards.gov.au/publications/Pages/20thaustraliantotaldietsurveyjanuary2003/20thaustralian totaldietsurveyfullreport/Default.aspx.

Gasnier, Céline, Coralie Dumont, Nora Benachour, Emilie Clair, Marie-Christine Chagnon, and Gilles-Éric Séralini. "Glyphosate-Based Herbicides are Toxic and Endocrine Disruptors in Human Cell Lines." *Toxicology* 262 (2009): 184–91.

"Glyphosate General Fact Sheet." National Pesticide Information Center, September 2010, http://npic.orst.edu/factsheets/glyphogen.pdf (accessed January 26, 2014).

Grandjean, Philippe, and Philip J. Landrigan. "Neurobehavioural Effects of Developmental Toxicity." *The Lancet Neurology* 13, no. 3 (March 2014): 330–38.

Guerrero-Bosagna, Carlos, Trevor R. Covert, Matthew Settles, Matthew D. Anway, and Michael K. Skinner. "Epigenetic Transgenerational Inheritance of Vinclozolin Induced Mouse Adult Onset Disease and Associated Sperm Epigenome Biomarkers." *Reproductive Toxicology* 34, no. 4 (December 2012): 694–707.

Guillette, Elizabeth A., Maria Mercedes Meza, Maria Guadalupe Aquilar, Alma Delia Soto, and Idalia Enedina Garcia. "An Anthropological Approach to the Evaluation of Preschool Children Exposed to Pesticides in Mexico." *Environmental Health Perspectives* 106, no. 6 (June 1998): 347–53.

Hardell, Lennart, and Mikael Eriksson. "A Case-Control Study of Non-Hodgkin Lymphoma and Exposure to Pesticides." *Cancer* 85, no. 6 (March 15, 1999): 1353–60.

Harras, Angela, ed. *Cancer Rates and Risks*. 4th ed. Washington, D.C.: U.S. Department of Health and Human Services, Public Health Service, National Institutes of Health, 1996.

Hayes, Tyrone B., Atif Collins, Melissa Lee, Magdelena Mendoza, Nigel Noriega, A. Ali Stuart, and Aaron Vonk. "Hermaphroditic, Demasculinized Frogs after Exposure to the Herbicide Atrazine at Low Ecologically Relevant Doses." *Proceedings of the National Academy of Sciences* 99, no. 8 (April 2002): 5476–80.

Hayes, Tyrone B., Kelly Haston, Mable Tsui, Anhthu Hoang, Cathryn Haeffele, and Aaron Vonk. "Atrazine-Induced Hermaphroditism at 0.1 ppb in American Leopard Frogs (*Rana pipiens*): Laboratory and Field Evidence." *Environmental Health Perspectives* 111, no. 4 (April 2003): 569–75.

Heinemann, Jack A., Melanie Massaro, Dorien S. Coray, Sarah Zanon Agapito-Tenfen, and Jiajun Dale Wen. "Sustainability and Innovation in Staple Crop Production in the US Midwest." *International Journal of Agricultural Sustainability*, published online June 14, 2013, http://www.tandfon line.com/doi/full/10.1080/14735903.2013.806408#.Ut766dLnaUk.

Higgins, Margo. "Toxins Are in Most Americans' Blood, Study Finds." *Environmental News Network*, March 26, 2001.

Hine, Rachel, and Jules Pretty. *Organic Agriculture and Food Security in Africa.* United Nations Environment Programme, United Nations Conference on Trade and Development, September 2008, http://www.unctad.org/en/docs/ditcted200715_en.pdf.

Howdeshell, Kembra, Andrew K. Hotchkiss, Kristina A. Thayer, John G. Vandenbergh, and Frederick S. vom Saal. "Environmental Toxins: Exposure to Bisphenol A Advances Puberty." *Nature* 401 (October 1999): 762–64.

Howe, Christina M., Michael Berrill, Bruce D. Pauli, Caren C. Helbing, Kate Werry, and Nik Veldhoen. "Toxicity of Glyphosate-Based Pesticides to Four North American Frog Species." *Environmental Toxicology and Chemistry* 23, no. 8 (August 2004): 1928–38.

Immig, Jo. "A List of Australia's Most Dangerous Pesticides." National Toxics Network, July 2010, http://www.ntn.org.au/wp/wpcontent/uploads/2010/07/FINAL-A-list-of-Australias-most-dangerous-pesticides-v2.pdf.

Infopest. Queensland Department of Primary Industries and Fisheries, Ann Street, Brisbane, Queensland, Australia, 2004.

International Agency for Research on Cancer and the World Health Organization. "Latest World Cancer Statistics, Global Cancer Burden Rises to 14.1 Million New Cases in 2012: Marked Increase in Breast Cancers Must Be Addressed." Press release, December 12, 2013.

International Assessment of Agricultural Knowledge, Science and Technology for Development (Washington, D.C.: Island Press, 2008).

International Food Policy Research Institute, www.ifpri.org.

Jayasumana, Channa, Sarath Gunatilake, and Priyantha Senanayake. "Glyphosate, Hard Water and Nephrotoxic Metals: Are They the Culprits Behind the Epidemic of Chronic Kidney Disease of Unknown Etiology in Sri Lanka?" *International Journal of Environmental Research and Public Health* 11, no. 2 (February 2014): 2125–47.

Jayawardena, Uthpala A., Rupika S. Rajakaruna, Ayanthi N. Navaratne, and Priyanyi H. Amerasinghe. "Toxicity of Agrochemicals to Common Hourglass Tree Frog (*Polypedates cruciger*) in Acute and Chronic Exposure." *International Journal of Agriculture and Biology* 12 (2010): 641–48.

Jerschow, Elina, Aileen P. McGinn, Gabriel de Vos, Natalia Vernon, Sunit Jariwala, Golda Hudes, and David Rosenstreich. "Dichlorophenol-Containing Pesticides and Allergies: Results from the U.S. National Health and Nutrition Examination Survey 2005–2006." *Annals of Allergy, Asthma & Immunology* 109, no. 6 (December 2012): 420–25.

Krüger, Monika, Shehata, Awad Ali Shehata, Wieland Schrödl, and Arne Rodloff. "Glyphosate Suppresses the Antagonistic Effect of *Enterococcus* spp. on *Clostridium botulinum*." *Anaerobe* 20 (April 2013): 74–78.

LaSalle, Tim J., and Paul Hepperly. "Regenerative Organic Farming: A Solution to Global Warming." Rodale Institute, 2008.

Laetz, Cathy A., David H. Baldwin, Tracy K. Collier, Vincent Hebert, John D. Stark, and Nathaniel L. Scholz. "The Synergistic Toxicity of Pesticide Mixtures: Implications for Risk Assessment and the Conservation of Endangered Pacific Salmon." *Environmental Health Perspectives* 117, no. 3 (March 2009): 348–53.

Lajmanovich, Rafael C., M. T. Sandoval, and Paola M. Peltzer. "Induction of Mortality and Malformation in *Scinax nasicus* Tadpoles Exposed to Glyphosate Formulations." *Bulletin of Environmental Contamination Toxicology* 70, no. 3 (March 2003): 612–18.

Lotter, D. W., R. Seidel, and W. Liebhart. "The Performance of Organic and Conventional Cropping Systems in an Extreme Climate Year." *American Journal of Alternative Agriculture* 18, no. 3 (2003): 146–54, http://www.donlotter.net/lotter_ajaa_article.pdf.

Lu, Chensheng, Kathryn Toepel, Rene Irish, Richard A. Fenske, Dana B. Barr, and Roberto Bravo. "Organic Diets Significantly Lower Children's Dietary Exposure to Organophosphorus Pesticides." *Environmental Health Perspectives* 114, no. 2 (February 2006): 260–63.

Lushchak, Oleh V., Olha I. Kubraka, Janet M. Storey, Kenneth B. Storey, Volodymyr I. Lushchak. "Low Toxic Herbicide Roundup Induces Mild Oxidative Stress in Goldfish Tissues." *Chemosphere* 76, no. 7 (2009): 932–37.

Maeder, Paul, Andreas Fliessbach, David Dubois, Lucie Gunst, Padruot Fried, and Urs Niggli. "Soil Fertility and Biodiversity in Organic Farming." *Science* 296 (May 2002): 1694–97, http://www.sciencemag.org/content /296/5573/1694.short.

Manikkam, Mohan, Carlos Guerrero-Bosagna, Rebecca Tracey, Md. M. Haque, and Michael K. Skinner. "Transgenerational Actions of Environmental Compounds on Reproductive Disease and Identification of Epigenetic Biomarkers of Ancestral Exposures." *PLoS ONE* 7, no. 2 (February 2012).

Manikkam, Mohan, Rebecca Tracey, Carlos Guerrero-Bosagna, and Michael K. Skinner. "Pesticide and Insect Repellent Mixture Permethrin and DEET Induces Epigenetic Transgenerational Inheritance of Disease and Sperm Epimutations." *Journal of Reproductive Toxicology* 34, no. 4 (December 2012): 708–19.

Marc, Julie, Odile Mulner-Lorillon, and Robert Bellé. "Glyphosate-Based Pesticides Affect Cell Cycle Regulation." *Biology of the Cell* 96, no.3 (April 2004): 245–49.

Mesnage, Robin, Benoît Bernay, and Gilles-Éric Séralini. "Ethoxylated Adjuvants of Glyphosate-Based Herbicides are Active Principles of Human Cell Toxicity." *Toxicology* 313, nos. 2–3 (November 2013): 122–28. Published online September 21, 2012, http://dx.doi.org/10.1016/j.tox.2012.09.006.

Mesnage, Robin, Emilie Clair, Steeve Gress, C. Then, A. Székácsd, and Gilles-Éric Séralini. "Cytotoxicity on Human Cells of Cry1Ab and Cry1Ac Bt Insecticidal Toxins Alone or with a Glyphosate-Based Herbicide." *Journal of Applied Toxicology* 33, no. 7 (July 2013): 695–99. Originally published online February 2012.

Mesnage, Robin, Nicolas Defarge, Joël Spiroux de Vendômois, and Gilles-Éric Séralini. "Major Pesticides are More Toxic to Human Cells than Their Declared Active Principles." *BioMed Research International* (2013), http://www.hindawi.com/journals/bmri/aip/179691/.

Millennium Ecosystem Assessment Synthesis Report. United Nations Environment Programme, March 2005.

Monbiot, George. "Organic Farming Will Feed the World." *Guardian*, August 24, 2000.

National Institute of Environmental Health Sciences, http://www.niehs.nih.gov. Accessed July 15, 2013.

Newbold, Retha R., Elizabeth Padilla-Banks, Ryan J. Snyder, and Wendy N. Jefferson. "Developmental Exposure to Estrogenic Compounds and Obesity." *Birth Defects Research Part A: Clinical and Molecular Teratology* 73, no. 7 (2005): 478–480.

Niggli, Urs. "Sustainability of Organic Food Production: Challenges and Innovations." *Proceedings of the Nutrition Society*, forthcoming.

Nilsson, Eric, Ginger Larsen, Mohan Manikkam, Carlos Guerrero-Bosagna, Marina I. Savenkova, and Michael K. Skinner. "Environmentally Induced Epigenetic Transgenerational Inheritance of Ovarian Disease." *PLoS ONE* 7, no. 5 (May 2012): e36129. doi:10.1371/journal.pone.00361.

Paganelli, Alejandra, Victoria Gnazzo, Helena Acosta, Silvia L. López, and Andrés E. Carrasco. "Glyphosate-Based Herbicides Produce Teratogenic Effects on Vertebrates by Impairing Retinoic Acid Signaling." *Chemical Research in Toxicology* 23, no. 10 (August 2010): 1586–95.

Parrott, Nicholas, and Terry Marsden. *The Real Green Revolution: Organic and Agroecological Farming in the South*. Canonbury Villas, London: Greenpeace Environmental Trust, 2002.

Pastor, Patricia N., and Cynthia A. Reuben. "Diagnosed Attention Deficit Hyperactivity Disorder and Learning Disability: United States, 2004–2006. National Center for Health Statistics. *Vital and Health Statistics* 10, no. 237 (July 2008).

Pearce, Fred, and Debora Mackenzie. "It's Raining Pesticides." *New Scientist*, April 3, 1999.

Pimentel, David, Paul Hepperly, James Hanson, David Douds, and Rita Seidel. "Environmental, Energetic and Economic Comparisons of Organic and Conventional Farming Systems." *Bioscience* 55, no. 7 (July 2005): 573–82, http://www.ce.cmu.edu/~gdrg/readings/2007/02/20/Pimental_EnvironmentalEnergeticAndEconomicComparisonsOfOrganicAndConventionalFarmingSystems.pdf.

Porter, Warren P., James W. Jaeger, and Ian H. Carlson. "Endocrine, Immune and Behavioral Effects of Aldicarb (Carbamate), Atrazine (Triazine) and Nitrate (Fertilizer) Mixtures at Groundwater Concentrations." *Toxicology and Industrial Health* 15 (January 1999): 133–50.

Qiao, Dan, Frederic Seidler, and Theodore Slotkin. "Developmental Neurotoxicity of Chlorpyrifos Modeled In Vitro: Comparative Effects of Metabolites and Other Cholinesterase Inhibitors on DNA Synthesis in PC12 and C6 Cells." *Environmental Health Perspectives* 109, no. 9 (September 2001): 909–13.

Rauh, Virginia, Srikesh Arunajadai, Megan Horton, Frederica Perera, Lori Hoepner, Dana B. Barr, and Robin Whyatt. "7-Year Neurodevelopmental Scores and Prenatal Exposure to Chlorpyrifos, a Common Agricultural Insecticide." *Environmental Health Perspectives* 119 (2011): 1196–1201. Published online April 21, 2011.

Rauh, Virginia, Frederica P. Perera, Megan K. Horton, Robin M. Whyatt, Ravi Bansal, Xuejun Hao, Jun Liu, Dana Boyd Barr, Theodore A. Slotkin, and Bradley S. Peterson. "Brain Anomalies in Children Exposed Prenatally to a Common Organophosphate Pesticide." *Proceedings of the National Academy of Sciences of the United States of America* 109, no. 20 (May 2012), www.pnas.org/cgi/doi/10.1073/pnas.1203396109.

Reganold, John P., Jerry D. Glover, Preston K. Andrews, and Herbert R. Hinman. "Sustainability of Three Apple Production Systems." *Nature* 410 (April 2001): 926–30.

Relyea, Rick A. "New Effects of Roundup on Amphibians: Predators Reduce Herbicide Mortality; Herbicides Induce Antipredator Morphology." *Ecological Applications* 22 (2012): 634–47.

Richard, Sophie, Safa Moslemi, Herbert Sipahutar, Nora Benachour, and Gilles-Éric Serálini. "Differential Effects of Glyphosate and Roundup on Human Placental Cells and Aromatase." *Environmental Health Perspectives* 113, no. 6 (June 2005): 716–20. Published online February 25, 2005, http://www.ncbi.nlm.nih.gov/pmc/articles/PMC1257596/.

Rodale Institute. "Organic No-Till," http://www.rodaleinstitute.org/no-till_revolution (accessed January 2014).

Samsel, Anthony, and Stephanie Seneff. "Glyphosate's Suppression of Cytochrome P450 Enzymes and Amino Acid Biosynthesis by the Gut Microbiome: Pathways to Modern Diseases." *Entropy* 15, no. 4 (2013): 1416–63.

————. "Glyphosate, Pathways to Modern Diseases II: Celiac Sprue and Gluten Intolerance." *Interdisciplinary Toxicology* 6 no. 4 (2013): 159–84 http://sustainablepulse.com/wp-content/uploads/2014/02/Glyphosate_II_ Samsel-Seneff.pdf (accessed March 21, 2014).

Séralini, Gilles-Éric, Dominique Cellier, Joël Spiroux de Vendômois. "New Analysis of a Rat Feeding Study with a Genetically Modified Maize Reveals Signs of Hepatorenal Toxicity." *Archives Environmental Contamination Toxicology* 52 (2007): 596–602.

Séralini, Gilles-Éric, Robin Mesnage, Emilie Clair, Steeve Gress, Joël Spiroux de Vendômois, and Dominique Cellier. "Genetically Modified Crops Safety Assessments: Present Limits and Possible Improvements." *Environmental Sciences Europe* 23, no. 10 (2011).

Seufert, Verena, Navin Ramankutty, and Jonathan A. Foley. "Comparing the Yields of Organic and Conventional Agriculture." *Nature* 485 (May 2012): 229–32, http://www.nature.com/nature/journal/v485/n7397/full/ nature11069.html.

Shehata, Awad Ali, Wieland Schrödl, Alaa A. Aldin, Hafez M. Hafez, and Monika Krüger. "The Effect of Glyphosate on Potential Pathogens and Beneficial Members of Poultry Microbiota in Vitro." *Current Microbiology* 66, no. 4 (2012): 350–58.

Short, Kate. *Quick Poison, Slow Poison: Pesticide Risk in the Lucky Country.* St. Albans, NSW: K. Short, 1994.

Skakkebæk, N. E., E. Rajpert-De Meyts, and K. M. Main. "Testicular Dysgenesis Syndrome: An Increasingly Common Developmental Disorder with Environmental Aspects." *Human Reproduction* 16, no. 5 (2001): 972–78.

Sparling, D. W., Gary Fellers. "Comparative Toxicity of Chlorpyrifos, Diazinon, Malathion and Their Oxon Derivatives to Larval *Rana boylii.*" *Environmental Pollution* 147 (2007): 535–39.

Storrs, Sara I., and Joseph M. Kiesecker. "Survivorship Patterns of Larval Amphibians Exposed to Low Concentrations of Atrazine." *Environmental Health Perspectives* 112, no. 10 (July 2004): 1054–57.

Swanson, Nancy. "Genetically Modified Organisms and the Deterioration of Health in the United States." *Sustainable Pulse*, April 27, 2013, http:// sustainablepulse.com/2013/04/27/dr-swanson-gmos-and-roundup-increase -chronic-diseases-infertility-and-birth-defects (accessed August 24, 2013).

Thongprakaisang, Siriporn, Apinya Thiantanawat, Nuchanart Rangkadilok, Tawit Suriyo, and Jutamaad Satayavivad. "Glyphosate Induces Human Breast Cancer Cells Growth via Estrogen Receptors." *Food and Chemical Toxicology* 59 (September 2013): 129–36, http://dx.doi.org/10.1016/ j.fct.2013.05.057.

Turnbull, Bob. "Research Shows Organic Corn, Soybean Yields Can Exceed Conventional." *Organic & Non-GMO Report*, January 2010, http://www .non-gmoreport.com/articles/feb10/organic_corn_soybean_yields_exceed_ conventional.php.

"U.S. President's Cancer Panel 2008–2009 Annual Report; Reducing Environmental Cancer Risk: What We Can Do Now." Suzanne H. Reuben for the President's Cancer Panel, U.S. Department Of Health And Human Services, National Institutes of Health, National Cancer Institute, April 2010.

Vandenberg, Laura N., Theo Colborn, Tyrone B. Hayes, Jerrold J. Heindel, David R. Jacobs Jr., Duk-Hee Lee, Toshi Shioda, Ana M. Soto, Frederick S. vom Saal, Wade V. Welshons, R. Thomas Zoeller, and John Peterson Myers. "Hormones and Endocrine-Disrupting Chemicals: Low-Dose Effects and Nonmonotonic Dose Responses." *Endocrine Reviews* 33, no. 3 (June 2012): 378–455. First published ahead of print March 14, 2012, as doi:10.1210/ er.2011-1050 (*Endocrine Reviews* 33: 0000–0000, 2012).

Welsh, Rick. "The Economics of Organic Grain and Soybean Production in the Midwestern United States." Policy Studies Report No. 13, Henry A. Wallace Institute for Sustainable Agriculture, May 1999.

INDEX

persistence of in the
environment, xxii, 2, 4–5,
6, 7, 49, 54–55, 56, 68,
118
usage in the developing
world, 57–58, 88, 92, 97
polycystic ovarian disease, 6, 23
precautionary principle, 62
push-pull system, 103–6

REACH, 60–61
Registration, Evaluation, and
Authorization of
Chemicals. *See* REACH
regulatory authorities
inconsistencies between,
81–82
limitations of, xxii, 1, 2, 6–7,
10, 25, 47, 53, 56,
57–60, 61, 82–83, 111,
117–20
need for revised chemical
limits, 21, 29, 30
reactionary versus
precautionary approach
of, 10, 60, 82, 115, 120
reliance on manufacturer-
sponsored studies, 59,
62–63, 71, 77–80, 113–14,
116–17
See also names of specific
organizations
Research Institute of Organic
Agriculture, 96
See also FiBL DOK Trials
residue levels, xxii, 30, 42, 43, 56,
67, 81, 119

retinoic acid signaling pathway,
65–66, 68–69
Rodale Institute, 89–90, 91
Farming Systems Trial, 90,
93–94
Roundup, 7, 8, 11, 12, 64, 65, 66,
69, 70–71
See also glyphosate

Silent Spring (Carson), xx
smallholder farms, 92, 96–99,
103, 104, 120
State of the Science of Endocrine
Disrupting Chemicals
2012, xx, 3
Swanson, Nancy, 36–37, 71–77
synergistic effects, 2–3, 4, 5, 12,
68, 118

thyroid
cancer, 37, 73, 113
disruption, 5, 34–35, 59
Tigray Project, 99–101
toxicology, 13, 16, 30, 38, 45
type 2 diabetes. *See under*
diabetes

U.S. President's Cancer Panel, 1,
13, 25, 62
2010 report, xx, xxi–xxii, 1, 2,
8, 13–15, 45, 50, 60, 82
United Kingdom, 91, 92, 103
United Nations Environment
Programme, 3, 25, 30,
33–34, 35, 37, 40–41, 46,
54, 59, 96, 97